"Lively and instructive . . . We experience [Begos s] discoveries on the wine side roads as freshly as he does." —*The Wall Street Journal*

"Kevin Begos jumped in, with the zeal of a seasoned journalist, after he was entranced by an Israeli wine he found in his hotel room minibar while reporting from the Middle East . . . He weaves a fascinating story that mixes personal exploration with cultural enlightenment. By searching for the origins of wine, Begos brings its past to us in the here and now. Along the way, he helps us understand how that wine we tasted on a long-ago vacation still resonates in our memory, our palate and our cultural identity." —*The Washington Post*

"Nothing less than the exact book the wine world needs right now." —*Toronto Star*

"His story unfolds in a manner similar to the growth of ancient grapevines; rooted in a strong central narrative, side stories grow like tendrils, wrapping around and supporting each other, while clusters of vividly described wines emerge like ripe grapes. Anyone who is interested in wine history, viniculture, or just enjoying a glass of wine will likely find *Tasting the Past* a pleasurable read." —*Science*

"Reads more like a detective tale and travelogue than a history lesson. [Begos's] skills as a reporter and a storyteller shine in this book." —*Forbes*

"With a cast of characters that includes archaeologists, botanists and historians, Kevin Begos offers a lively account of his journey to find the origins of wine, debunking a number of the wine world's long-held beliefs along the way." —*PUNCHdrink.com*

"If you can tell Sauvignon blanc from Sémillon, you might feel that you 'know' wine. Science journalist Kevin Begos blows that idea to smithereens. He traveled from the Caucasus Mountains to Israel and beyond, and riffled through archives, to unearth ancient 'founder' grape varieties. En route, he consults archaeobiologist Patrick McGovern and grape geneticist Shivi Drori; reads papers on the DNA of 'wild yeasts that live symbiotically with wasps'; and contemplates the oldest grape fossil found. A book that froths with data on half-forgotten vines, from Hamdani to Gros Manseng."   —*Nature*

"Begos does an excellent job striking a balance between travel writing, history, and science . . . There are so many great stories and characters contained within *Tasting the Past*'s 250 or so pages, and Begos's journalistic style keeps it all moving. Anyone who wants to know what else is out there, beyond even what your local Total Wine can supply, will want to read this book. Those with a bent toward wine history, paleobotany, or grape genetics will be especially pleased."   —*Terroirist: A Daily Wine Blog*

"The fascinating culmination of ten years of research into the birth of wine, and the uncovering of all sorts of forgotten grapes."
                                                            —*BuzzFeed News*

"Oenophiles will raise a glass to Begos's excellent exploration of the science and history of wine, now a $300 billion global market . . . This mix of memoir and wine education guide is all-around satisfying."
                                                            —*Publishers Weekly*

"Readable and well researched . . . A palatable blend of history, food science, and travel writing."   —*Library Journal*

"Kevin Begos has pulled together a wonderfully surprising assortment of characters and threads to take us on his compelling journey into wine's origin."   —Alice Feiring, author of *The Dirty Guide to Wine*

"If you are a fan of the food writing of Mark Kurlansky or Michael Pollan, for example, you probably want to have this book on your radar. If you like esoteric wine, ditto. If you like the idea that wine gives us a way of drinking history? That too. This is a smart book, and of every title on this list the one I most wish I'd written."   —*Paste Magazine*

"Kevin Begos's exploration of the world's storied wines is an intoxicating blend of scientific inquiry, human culture, and natural history. Pull up a chair, open the book and pour yourself a glass to go with it—you'll enjoy it more than ever."

—Deborah Blum, author of *The Poisoner's Handbook*

"Fascinating . . . An intriguing tale of discovery, setting scenes and meeting with scientists, historians, winemakers, scholars, enologists, and private enthusiasts, among others, to discover where wine was first fermented, what ancient wine really tasted like, and what's being done to identify and preserve ancient grape varieties around the globe."

—*Spirited* magazine

"[Begos] consults fascinating characters, including a biomolecular archaeologist, master sommeliers, and vintners striving to save their local culture by producing traditional wines from days gone by. As Begos continues his search, it becomes clear to him, as well as to readers, that getting to know wine-producing regions' people, cultures, and history is integral to truly understanding what they've produced—and are yet to."   —*Booklist*

"This is quite a book and I hope it is read widely throughout the wine world and that it has a huge impact. The fact that current practices have put a halt to evolution for wine grapes, that was news to me. *Tasting the Past* shocked the hell out of me."

—Kermit Lynch, wine merchant and author of
Adventures on the Wine Route

"Investigates the variety of wine grapes, as part of an engaging personal account, ten years in the making."

—*The Times Literary Supplement* (UK)

"Insightful. This well-researched book has combined history, science, and travel in one place, which makes it extraordinarily interesting and entertaining. It is packed with drink recommendations and will teach you how to select your wine the next time you go out to buy a bottle. This intoxicating and sensual book will change the way you think of wine. *Tasting the Past* is an essential read for every wine lover."

—*The Washington Book Review*

"A vintner's blend of science, history, travel, and tantalizing drink recommendations. It's the kind of book that will have you quoting passages at cocktail parties, making off-the-beaten-path travel plans, and seeking out unusual and obscure wines everywhere you go."

—Amy Stewart, author of *The Drunken Botanist*

"From the moment Begos describes Cremisan wine, the reader is captivated. You don't have to be an oenophile to dig this book."

—*The Atlanta Journal-Constitution*

"In a personal, conversational style, Kevin Begos mixes poetry, history, geography, and science into a delightful volume in praise of unique, nearly forgotten grapes that give real meaning to terroir."

—Mark Pendergrast, author of *Uncommon Grounds: The History of Coffee and How It Transformed Our World*

"Ought to be a must-read for any self-respecting wine nerd."

—*Good Food Revolution*

"A myth-busting, history-reclaiming, science-centric, skeptical—and, yet, loving and respectful—tour of the history, the present, and even the future of wine production. Kevin Begos is an unrelenting and delightful detective."

—Cat Warren, *New York Times* bestselling author of
*What the Dog Knows: Scent, Science, and the
Amazing Ways Dogs Perceive the World*

"Earthy, with undertones of humor. This fascinating dive into the world of obscure wines will educate you, entertain you, and make you want to drink. Begos teaches us that a glass of wine isn't just a glass of wine, but a window into history, culture, and science."

—Matti Friedman, author of *Pumpkinflowers*

"It just captivated me and never let go. Kevin Begos has created the most currently documented history of wine, as it exists, in the most masterful way possible. This wine book, just like wine, is here far from dry, and carries all of the excitement of a 1945 Margaux Bordeaux."

—*Wine Blog: Juicy Tales by Jo Diaz*

"[The] book is an example of the best journalism, which arises not from an event or, worse by far, an agenda, but from a question whose answer enriches all. Begos may very well encourage and hasten the wine diversity trend."

—*Tallahassee Magazine*

"Begos is an accomplished writer who enlightens with insights about the spread of winemaking throughout the world . . . and the joy of unexpected grape variety experiences that go beyond the established varieties such as Cabernet, Pinot Noir, Riesling and Chardonnay."

—*The Prince of Pinot*

"You don't need to be an oenophile to enjoy this flavorful adventure about one wine nerd's search for the perfect grape . . . This multidisciplinary master class in the history, science, religion, and literature of wines is as luscious as a full-bodied pinot noir."

—*Kirkus Reviews*

"The most fascinating wine book I've read in years."
—Dave Nershi, *Vino-Sphere*

"One of the most interesting books about wine I've ever read. It's Vin BC! Belongs in every oenophile's library."
—Eugenia Bone, author of *The Kitchen Ecosystem*

"Through mythology, history, flavor and science, *Tasting the Past* stretches across centuries and continents, helping us understand why the transformation of juice to alcohol continues to intoxicate us all. A compelling, sensual journey weaving past and present."
—Simran Sethi, author of *Bread, Wine, Chocolate*

TASTING THE PAST

TASTING THE PAST

One Man's Quest to Discover (and Drink!)
the World's Original Wines

Kevin Begos

ALGONQUIN BOOKS OF CHAPEL HILL   2019

Published by
Algonquin Books of Chapel Hill
Post Office Box 2225
Chapel Hill, North Carolina 27515-2225

a division of
Workman Publishing
225 Varick Street
New York, New York 10014

Lines from an Abu Nuwas poem are from Roger M. A. Allen,
Encyclopedia Britannica online, "Arabic Literature."

Lines from a poem by Israel ben Moses Najara are from Israel Zinberg, *A History of Jewish Literature*, vol. 5, *The Jewish Center of Culture in the Ottoman Empire*, Bernard Martin, translator.

The Richard Feynman comments are from *The Feynman Lectures on Physics*, vol. 11,
The Complete Audio Collection (New York: Basic Books, 2007).

Lines from A. Leo Oppenheim, *Letters from Mesopotamia* (1967) are used with
permission from The Oriental Institute of the University of Chicago.

Translations and an illustration from Eva-Lena Wahlberg's research
on Egyptian wine labels are used with her permission.

Lines from Andrew George's translation of *The Epic of Gilgamesh* are used
with his permission.

Nart poetry is from *Nart Sagas from the Caucasus: Myths and Legends
from the Circassians, Abazas, Abkhaz, and Ubykhs*, John Colarusso, translator
(Princeton University Press, 2002).

Lines from a poem by Samuel the Nagid are from *A Miniature Anthology of
Medieval Hebrew Wine Songs*, Raymond P. Scheindlin, translator.

The Library of Congress has catalogued the hardcover edition as follows:

Library of Congress Cataloging-in-Publication Data
Names: Begos, Kevin (Kevin Paul), author.
Title: Tasting the past : the science of flavor and the search for the original
wine grapes / Kevin Begos.
Description: First edition. | Chapel Hill, North Carolina : Algonquin Books
of Chapel Hill, 2018. | Includes bibliographical references.
Identifiers: LCCN 2017046783 (print) | LCCN 2017055055 (ebook) |
ISBN 9781616208233 (ebook) | ISBN 9781616205775 (hardcover : alk. paper)
Subjects: LCSH: Grapes—Varieties. | Grapes—History. | Viticulture. |
Wine and wine making—History.
Classification: LCC SB398.28 (ebook) | LCC SB398.28 .B44 2018 (print) | DDC 634/.83—dc23
LC record available at https://lccn.loc.gov/2017046783

ISBN 978-1-61620-937-7 (PB)

10 9 8 7 6 5 4 3 2 1
First Paperback Edition

For my mother and father,
two book lovers

CONTENTS

PART ONE

We're still caught in that trap of saying, well, there are only ten good [grape] varieties in the whole world, and that's it. There are wonderful wines to be made everywhere from a huge number of varieties.

—ANDY WALKER, LOUISE ROSSI ENDOWED CHAIR IN VITICULTURE, UNIVERSITY OF CALIFORNIA, DAVIS

1

A Mysterious Wine

The Wine of Life keeps oozing drop by drop,
The Leaves of Life keep falling one by one.
—OMAR KHAYYAM, *THE RUBAYIAT*, CA. 1100

Alone in Amman, Jordan, I looked at the mini-bar skeptically yet wistfully. Finding good wine in a hotel room is a tantalizing concept, but I had a rule: never buy the stuff. This place had rustic tiles and carved wooden doors in the lobby that gave way to generic rooms—clean and fairly comfortable, but like a Holiday Inn, without character. I was fidgety. Watching Arabic TV without understanding a word was only briefly entertaining. The evening call to prayers from a nearby mosque reminded me of the limited alcohol options, and I knew hardly anyone in the city. I went over to the TV cabinet, opened the door again, and sadly contemplated the row of bottles next to the little refrigerator. One red wine had an unusual label with old-fashioned type and images. It read:

Produced and Bottled by Cremisan Cellars
HOLY LAND—Bethlehem

That seemed odd. It was the spring of 2008, and there were still vineyards in Bethlehem? My hazy Catholic childhood taught me that people drank wine there in biblical times, but I'd never seen Cremisan on a store shelf or restaurant list, or in a review. The label said they started making wine in 1885, which I found interesting but also curious. Had no critics checked it out? The winery is just a few miles from Jerusalem.

The bottle was the only thought-provoking thing in the room and I was tired, physically and emotionally, from a Middle East reporting assignment. With low expectations I broke my rule, pulled the cork, and took a sip. *Wow*. I perked up immediately. The dry red wine had a spicy flavor, sort of Syrah-ish, but not quite. It was drinkable, balanced, and pleasingly different, with even a hint of earthy terroir. I went to bed happier, imagining it might be fun to visit Cremisan.

But deadlines beckoned for the rest of the trip. I returned to the United States planning to buy another bottle at home to share with friends. No luck. Visits to wine stores drew blank stares. Cremisan didn't have an American importer, so I couldn't even buy the wine online. When I told the story, some suggested that I airfreight a few bottles. They didn't get the torturous nature of international wine shipping laws. Bethlehem is technically in the Palestinian Territories, and that didn't help either, especially after the Second Intifada, or uprising. I tried emailing the winery but got no response.

There was something else, too. I was getting bored with the Chardonnays, Merlots, Cabernet Sauvignons, and Rieslings in every wine store and on every restaurant list. I'm not against those grapes—of course they make some wonderful wines. But why the lack of diversity? Wasn't there more?

Little details about Cremisan kept me intrigued. Once, when I looked for news of my wine-madeleine, I found Cremisan's rudimentary website. The winemaker was an Italian monk, and the monastery was built near the ruins of a seventh century Byzantine church. They made some Merlot, but also used local grapes I'd never heard of: red Baladi, white Jandali, and Hamdani. Had they really grown in this region for thousands of years? Did Cremisan use native grapes to make wine that the Egyptians, biblical prophets, and Romans might have drunk?

The buying difficulty was perplexing. Why wouldn't a nice, modestly priced wine from a vineyard in the heart of the Holy Land have at least a small built-in market? Perhaps the monks were indifferent to worldly marketing. Whoever heard of monks in the Middle East making wine in the twenty-first century, anyway?

The website showed various labels with a jumble of multicultural wine references: Côtes de Cremisan, Old Hock, David's Tower, Cana of Galilee, and Blanc de Blanc, "made from selected Daboky local white grapes grown in the mountains of Bethlehem." I listed influences and came up with Italian–French–Christian–Jewish–Arabic–British–German–Spanish. In a region so fractured by ethnic and religious wars, this melting pot was sort of endearing.

The memory of my hotel room wine lingered like the refrain of an old pop song. In 2011 I moved on to a busy job as a correspondent for the Associated Press, and soon afterwards I looked for Cremisan or their mysterious grapes in the third edition of the authoritative *Oxford Companion to Wine*, which was the current one at the time. Nothing. The entry on Israel said vineyards there were destroyed after the spread of Islam in AD 636, and that "the Crusaders temporarily restored wine production between AD 1100 and 1300, but with the exile of the Jews, vine-growing ceased." Supposedly no real wine industry existed until Baron Edmond de Rothschild started a winery in the 1880s. Even worse, it said Israel currently had "no indigenous varieties" of grapes.

But another entry on a nearby page of the *Oxford Companion* [under *Islam*] said that in the Middle Ages "Muslim conquest by no means outlawed wine production" and that Arabs even tried finer wines made in the Christian monasteries of Iraq. Abu Nuwas, who lived in Baghdad around AD 800, was famous for his scandalous poems about drinking, with lines such as *"[P]our me a glass of wine, and confirm that it's wine! / Do not do it in secret, when it can be done in the open."*

I was confused. Wine's supposed disappearance from the Holy Land seemed to be one of those lazy cultural stereotypes that appears to make perfect sense, until you realize it doesn't. That region had significant Christian and Jewish communities from the seventh to twentieth centuries, under many different rulers. Islam's wine prohibition didn't apply to them. How could winemaking and vineyards have vanished?

But I didn't really have the time to explore Middle Eastern wine history. Tracking down bottles from Cremisan fell to the bottom of my to-do lists. I'd stumbled on an unusual wine, nothing more. From time to time, though, whenever I read a little pocket-sized book of Persian poetry I was fond of, I'd be reminded of possible holes in the *Oxford Companion* narrative. In the 1200s Rumi wrote: *That jug full of wine has brought me to such a state that I have broken so many jars today.*

Sometimes the Muslim wine poets spoke metaphorically about alcohol, but to me the vivid lines also suggest real-life experience. Medieval Jewish wine poets who had lived in what is now Israel wrote about wine, too.

> When your heart is cradled in sorrow
> and trouble today and tomorrow,
> day after day,
> drink the liquid of the grape, my friend
>
> —ISRAEL BEN MOSES NAJARA (CA. 1555–1625)

Persians, Arabs, and Jews, all reciting wine poems through the centuries when vineyards supposedly disappeared because of Islam? But poems aren't historical proof, and who was I to question *The Oxford Companion to Wine*, anyway? I could even see why the modern wine trade ignored such history, and Cremisan. As I pored over wine poems Chinese billionaires bought up Bordeaux vineyards, and California's Napa Valley boom continued, helped by surging tech fortunes and eager consumers. Single bottles from Burgundy's Domaine de la Romanée-Conti, one of the world's most famous wineries, auctioned for more than ten thousand dollars. Cremisan was a viticultural David up against Goliaths. Why care about the history of obscure Middle Eastern grapes when there is so much money sloshing around the wine world?

Memories of the hotel room bottle faded. I gave up looking for Cremisan wine. Until I heard about José Vouillamoz's work. He co-authored the landmark 2012 book *Wine Grapes* with Jancis Robinson and Julia Harding, and it raised my hopes. The authors described 1,368 varieties of grapes used to make wine all over the world, explored the history of each one, and included diagrams of wine grape family trees based on DNA research. The book won a James Beard Award plus many other international honors. It's an encyclopedia of famous varieties, plus rare ones on the verge of extinction. The book enthralled me, and exposed the extent of my wine ignorance. I eagerly typed "Cremisan" into the *Wine Grapes* ebook search box, along with the names of their unusual grapes.

Nothing.

I contacted Vouillamoz, hoping for some shred of encouragement. "I must confess something almost embarrassing. We missed them. The grapes they are growing are not in our book because at that time we had never heard of them—neither me nor Jancis Robinson. So it shows you how obscure they were, at least a few years ago," he told me. Robinson is one of the most famous, knowledgeable, and respected critics in the world. I was infatuated with a winery she'd never heard of. Maybe I'd found an arcane wine history error about who drank what when in the *Oxford Companion*. So what? It was all feeling like some wild grape chase, trying to link Cremisan to a perhaps mystical ancient wine.

But *Wine Grapes* led to a different realization: I wanted to learn about wine, not be preached at. I was tired of excessively florid reviews with laundry lists of all the flavors and aromas in a glass, and skeptical of critics who claimed the ability to distinguish between 92-, 95-, or 100-point bottles. Now I saw the obvious: wine grape DNA can be analyzed the way a home cheek-swab kit delivers data about your family tree. Evolution leaves a trail, whether it's Neanderthals splitting from *Homo sapiens* or the birth of a new grape variety. My hotel room wine morphed from an obsession into training wheels. I understood that obscure grapes weren't just quirkily interesting—each had a special flavor profile. Losing one could mean losing certain tastes forever.

Vouillamoz's genetic detective work inspired all sorts of questions: Where did wine grapes originate? I didn't even know how to define "ancient" grape varieties—did it mean five hundred years old, or five thousand? And why do grapes express so many flavors and aromas, when rice and grains, also used to make alcohol, don't? I had no idea.

Wine Grapes also suggested the beginnings of a movement: vineyards all over the world determined to preserve native grapes. I made lists of the curious wines from Armenia, Greece, France, and many other countries, wanting to try them all: Assyrtiko, Chinuri, Kisi, Maratheftiko, Nero d'Avola, Saperavi, Tannat, Violento, Xynisteri—the vineyard equivalents of Slow Food, multicolored heirloom vegetables, or craft whiskey. For the first time I understood that endless vineyards of Cabernet, Chardonnay, and Merlot had pushed aside hundreds of local varieties over the last century.

Other connections emerged. The modern wine industry looked more and more like a dressed-up version of industrial agriculture, designed to produce maximum yields at the lowest cost. That reminded me of Michael Pollan's observation about apples in *The Botany of Desire*: "[T]he industry got together and decided it would be wise to simplify that market by planting and promoting only a small handful of brand-name varieties." Thomas Jefferson loved Esopus Spitzenburg apples, and early Americans enjoyed dozens more: Newtown Pippins, Roxbury Russets, and Ashmead's Kernels. But in the twentieth century Red Delicious, Golden Delicious, Granny Smith, and McIntosh took over in much the same way a few French grapes took over the world's vineyards. It's not just apples and wine. Take bananas. One called Cavendish occupies 90 percent of the world market, even though there are about a thousand banana varieties in the world. Flowers? I knew from Amy Stewart's *Flower Confidential* that most popular varieties have almost no scent, because the industry cares more about size, color, and the ability to travel.

Corporate winemaking even helped limit some wines made from the famous grapes. The *Atlantic* published the article "The Dark Side of Wine" by wildly influential wine critic Robert M. Parker Jr. in 2000. In it he opined: "[I]t seems to be the tragedy of modern winemaking that it is now increasingly difficult to tell an Italian Chardonnay from one made in France or California or Australia. When the corporate winemakers of the world begin to make wines all in the same way, designing them to offend the least number of people, wine will no doubt lose its fascinating appeal and individualism to become no better than most brands of whiskey, gin, scotch, or vodka."

Statistics show there is cause to worry. A study in the *Journal of Wine Economics* found that between 1990 and 2010 Cabernet Sauvignon and Merlot more than doubled their share in the world's vineyards. By 2010 French grape varieties comprised 67 percent of vineyard acreage in New World countries, up from 53 percent just ten years before. Wine is supposedly all about variety, but it's easier and cheaper to limit the range of tastes and flavors we experience. At their best, the leading French grapes produce luscious, beautifully complex wine, but that's alongside a tidal wave of industrial, filtered, so-so product. In 2014 the Bordeaux region alone produced about seven hundred million bottles.

All this new information was overwhelming, and it raised questions for me. *Wine Grapes* illuminated the loss of wine grape diversity and the efforts to preserve rare varieties, but it was a work in progress. The big, foldout genealogy charts focused on Europe. There were gaps in the Caucasus Mountains and the Middle East, regions that were making wine thousands of years before the French or Italians. That seemed akin to using Columbus as the starting point for North American history.

Yet *Wine Grapes* did a superb job of showing how many European grapes share a few common ancestors. A once scorned grape called Gouais Blanc is related to more than eighty different varieties, including Riesling, one of the world's most popular grapes. The point? Even if you don't care about rare grapes, we wouldn't be drinking many wines without them.

So I was back to wondering: Were the varieties Cremisan used part of an earlier, forgotten network of "founder grapes"? The DNA of some undiscovered vine should have the answer, in theory. Perhaps I could seek out rare, native grapes with unusual tastes and try to understand the origins of wine at the same time. A scientific puzzle with wine tasting? Count me in.

One day I visited the *Wine Grapes* website and saw a note. The authors explained that they had missed some grape varieties and asked for suggestions, promising to include the first ten winners in future editions of the book. There they were: Cremisan's grapes. Gal Zohar, the sommelier for the world-famous London restaurant Ottolenghi, won for submitting the Baladi, Jandali, Dabouki, and Hamdani grapes. After five years of futile scrounging for information, I no longer felt so alone. Someone else liked the Cremisan wine, and soon Jancis Robinson was writing articles about it.

2

Archaeobiology and Ancient Wine

Beside the sea she lives, the woman of the vine, the maker of wine;
Siduri sits in the garden at the edge of the sea.
—The Epic of Gilgamesh

remisan and *Wine Grapes* led me to another scientist. University of Pennsylvania archaeologist Patrick McGovern is often called the Indiana Jones of ancient wine and beer. He received his undergraduate chemistry degree from Cornell and his neurochemistry PhD from the University of Rochester's Brain Research Center. He studied Biblical Archaeology in Jerusalem, lectured on Egypt in the time of the pharaohs at Rutgers, and has done fieldwork in Jordan, Syria, Iran, Armenia, China, and many other countries.

McGovern was one of José Vouillamoz's early mentors—in 2004 they traveled to the Middle East together to research wine grape origins. In the 1980s McGovern helped pioneer an emerging field: biomolecular archaeology. Even unbelievably faint residues (one part per million—or billion) on ancient pottery still contain chemical fingerprints of the liquids and foods they once held. Inspired by advances in radiocarbon dating and medical analysis, archaeologists and botanists use mass spectrometry and even nuclear reactors to decipher what people ate and drank thousands of years ago. "[A mass spectrometer] weighs molecules, sorts them according to weight, then counts the number of each weight," according to Harold Wiley, a pioneer who helped expand use of the tool in the 1940s. The molecular weight can reveal the exact compound that's present.

I sent McGovern an email about the Cremisan grapes; sadly, he didn't have anything to offer, leaving me with yet another of the world's leading wine experts who didn't seem excited about what I'd come to think of as "my" grapes. But McGovern's research was thought-provoking. It sometimes reads like a fictional script for *CSI: Ancient Alcohol*, or proof of physicist Richard Feynman's famous comment about wine's complexity.

Feynman worked on the Manhattan Project in World War II and won a Nobel Prize for quantum electrodynamics research— how matter and light interact. During a lecture in the 1960s he made these observations:

> A poet once said, "The whole universe is in a glass of wine."
> I don't think we'll ever know in what sense he meant that,
> for the poets don't write to be understood. But it is true that
> if you look at a glass of wine closely enough you will see the
> entire universe. There are the things of physics: the twisting liquid, the reflections in the glass, and our imagination
> adds the atoms. It evaporates depending on the wind and
> weather. The glass is a distillation of the earth's rocks, and

in its composition, as we've seen, the secret of the universe's age, and the evolution of the stars. What strange array of chemicals are in wine? How did they come to be? There are the ferments, the enzymes, the substrates, and products. And there in wine is found the great generalization: all life is fermentation. [. . .] If our small minds, for some convenience, divide this glass of wine, this universe, into parts— the physics, biology, geology, astronomy, psychology, and all—remember that nature doesn't know it. So we should put it all back together, and not forget at last what it's for. Let it give us one final pleasure more: drink it up and forget about it all.

Processes such as liquid chromatography (Greek word root: *chroma* [color] + *graphia* [drawing/writing]) revolutionized the study of ancient wine. It can separate a sample of just a few parts per trillion into distinct chemicals. Picture deconstructing a recipe down to the molecular level, or cleaning a tiny fleck of dried something-or-other from your refrigerator, and reverse-engineering it to uncover the original food. It works because the molecules of different elements vary in weight. Liquid chromatography puts tiny samples in liquid form, where the various elements move at different speeds through a test container of solid material, thus precisely isolating the compounds for further study. Such tests helped identify the types of wine in King Tut's tomb. Spanish researchers identified three types made from about 1500 to 1075 BC: a red, a white, "and a more elaborate red wine, called *shedeh*."

One of McGovern's projects looked at residue on five-thousand-year-old Egyptian amphoras—the clay containers that carried wine, oil, fruit, and grain all over the Mediterranean—and compared them to an amphora from about AD 500. The older samples came from the tomb of one of Egypt's first rulers, Scorpion I, which contained about seven hundred amphoras in three chambers.

The first challenge was proving what the contents were—it could have been wine, or not. However, the scientists recovered a tiny but identifiable fragment of *Saccharomyces cerevisiae* DNA—yeast—and tartaric acid, which is a grape biomarker. Some containers also had remains of grape seeds and of single figs sliced in half, perhaps to add sweetness or improve fermentation. The Scorpion I wine showed possible traces of coriander, wormwood, blue tansy, and pine resin, which was used as a combination flavoring/antioxidant. The chemical residue could also have come from mint, sage, or thyme.

The remains made sense when McGovern compared them to Egyptian papyrus writings about medicine, since herbs and spices served as flavorings and drugs for the ancient world. One papyrus is over a hundred pages long, and many of the thousand-plus prescriptions were mixed with wine and beer—likely an effective way to get medicine to patients, and early evidence of homeopathy. In recent years various researchers in the Middle East and throughout the Mediterranean have found evidence for a wide range of ancient wine flavorings: frankincense, myrrh, cumin, dill, fennel, aloe, and balm. At another Egyptian site dating to about AD 400, amphoras were scattered around Nubian-era taverns like kegs at a biker bar, suggesting wine had become a drink for commoners. Some of the people were buried nearby with amphoras, too. The typical size held thirty or forty gallons.

Amphoras are found all over the Mediterranean, and one huge Egyptian dumping ground may contain remains of more than a million. Not too long ago scientists counted, measured, and photographed them. To find out where they were manufactured, McGovern put some of the shards inside a nuclear reactor. Using instrumental neutron activation analysis (INAA), tiny, ground-up samples of four-thousand-year-old pottery were bombarded with neutron beams, so their basic chemical elements turned radioactive. The precise soil composition found on each sample was measured using gamma-ray emissions, providing a soil fingerprint. McGovern's team compared the results to databases of clay deposits. The result? Amphoras weren't made all over the place. Just as Pittsburgh was known for steel, and Detroit for cars, most of the ancient clay came from a narrow coastal region near the present-day Israeli city of Ashkelon.

McGovern thinks the people there specialized in the trade, making about five hundred jars per month over a two-hundred-and-fifty-year period. That research lends credence to a passage the Greek historian Herodotus wrote around 425 BC, describing one of the first recycling programs in history, which delivered used Egyptian wine jars to Syria:

Twice a year wine is brought into Egypt from every part of Greece, as well as from Phoenicia, in earthen jars . . . The burgomaster of each town has to collect the wine-jars within his district, and to carry them to Memphis, where they are all filled with water by the Memphians, who then convey them to this desert tract of Syria. And so it comes to pass that all the jars which enter Egypt year by year, and are there put up to sale, find their way into Syria, whither all the old jars have gone before them.

The Egyptians also wrote poems about wine dating back to at least three thousand years ago, such as this line from *The Flower Song*: "To hear your voice is pomegranate wine to me: I draw life from hearing it." That poem led me back even further, to a Sumerian drinking song from about four thousand years ago, sung along the Silk Road or in cities of the Fertile Crescent: "We are in a happy mood, our hearts are joyful! [. . .] Let the pouring of the sweet liquor resound pleasantly for you!" And this passage is from *The Tale of Sinuhe*, an epic Egyptian poem written roughly four thousand years ago:

> It was a good land called Yaa.
> Figs were in it and grapes.
> It had more wine than water.

My vague notions of ancient wine were expanding in all directions. I'd expected simplicity, yet found complexity and strong emotions from various cultures, much like the passionate sentiments of wine enthusiasts today. The Egyptians named vineyards, winemakers, and the quality of various vintages. I read early poems and stories from various cultures that distinguished between beer, fruit wine, grape wine, medicinal wine, and even sesame wine. I learned about Geštinanna, a Sumerian wine goddess, and Ninkasi, a goddess of alcohol and fermentation. Heroes and heroines went on epic journeys to netherworlds awash in plenty of intoxicating liquids. I even smiled at ancient jokes: "If I had a wine jar on my shoulder, my shoulder would not hurt."

One January I saw a notice for a Patrick McGovern lecture, and decided to go. "Uncorking the Past: Fermentation as Earth's Earliest Energy System and Humankind's First Biotechnology" promised to answer some of my basic wine science questions. About two hundred students, faculty, and members of the public filled the University of Alabama auditorium, which seemed like a sizeable turnout for a wonky lecture. After a short introduction McGovern pointed to a slide of *S. cerevisiae*—the yeast winemakers, brewers, and bakers use—and told the audience, "What it does is take sugar, and excrete it, you might say, into alcohol and also into carbon dioxide. This would have been quite exciting, I think, to ancient humans. It was like some sort of mysterious force from beyond that was causing this transformation." Fermentation helps preserve food, adds flavors, and produces a by-product with mind-altering effects, McGovern explained, so humans had plenty of reasons to worship the process, even if they didn't understand how it worked.

Animals imbibe, too, from fruit flies all the way up to elephants. "The fruit fly feeds its young with alcohol, which is very curious." McGovern said that the fly has many of the same genes for "getting drunk" as humans. Other researchers used a somewhat bizarre contraption called an inebriometer to precisely measure the tipsiness of individual fruit flies as they wobbled down test tubes, like some insect version of a state trooper asking Saturday night drivers to walk a line. They christened the relevant genes with joking nicknames:

- Lightweight: genes that make flies especially susceptible to alcohol

- Barfly: a mutation that allow flies to absorb a lot of alcohol without going to sleep

- Tipsy: genes that cause flies to conk out after just a little booze

McGovern told the crowd about birds feeding on fermented fruit and getting so drunk they fall off branches, and of zebra finches that slur their songs. The Malaysian tree shrew drinks the equivalent of nine glasses of wine a night in fermented palm nectar, yet doesn't get inebriated for some unknown reason. Elephants raiding Indian breweries have inspired headlines such as "Is Every Single Elephant a Village-Wrecking Booze Hound?" and "Trunk and Disorderly," though some question how much alcohol the animals really ingested.

I'd come to the lecture expecting a lot of talk about wine and beer, but McGovern made a broader point: a wide variety of species love or struggle with alcohol, all driven by individual genes. That suggests a primordial origin, perhaps as far back as 140 million years ago, when the first fruiting trees emerged. There's cosmic alcohol, too, McGovern said, such as ethyl formate in the Sagittarius B2 dust cloud. "It turns out that in the star-forming regions of the Milky Way . . . right at the center, maybe twenty-five million light years from here, there are huge clouds of alcohol. This has been established by spectroscopic methods. So we could go and mine some of that alcohol if we had the right kind of spaceship to get us out there," he joked. "It does show you that alcohol is an intrinsic compound in the universe."

McGovern said humans probably made fermented beverages at a very early date, meaning tens of thousands of years ago. But there's no evidence left to analyze. Pottery didn't exist for most of the Paleolithic Period / Old Stone Age era, which began 2.6 million years ago and concluded about 10,000 years ago with the end of the last ice age. According to McGovern, human sensory organs haven't changed much over time, so it's likely we explored ways to ferment alcohol long before the earliest evidence of Middle Eastern pottery, which is about only 8,000 years old. But pottery made the booze more storable and transportable.

Towards the end of the lecture an offhand comment from McGovern excited me. He spoke of using data from experimental archaeology in the real world, to re-create ancient beers. He said it was "absolutely necessary" to taste a wide variety of modern beverages to understand the old ones. The audience laughed, but I felt a kinship from my Cremisan experience.

In 1999 McGovern led a team that used archaeobiology to identify the food and drink served during King Midas's funeral feast, held in about 700 BC in what is now central Turkey. The scientists examined all the different residues on the bowls and jugs found at the site in order to re-create a sort of culinary time capsule. All the vessels held the same mixture of grape wine, barley beer, and honey mead, and eighteen jars found to contain food remains revealed the menu: a stew of goat or sheep meat, perhaps marinated with honey, wine, and olive oil, then cooked with lentils and spices.

After the lecture I introduced myself and McGovern remembered our email exchange about Cremisan. "We drank it when I lived on a kibbutz in Israel," he said of the 1970s, but without paying attention to the grape varieties.

McGovern's archaeological and laboratory research was enchanting, but I was interested in the vineyards and wines of the present, too. I'd visited the Middle East several times, mostly writing about conflict and turmoil. I was tired of all that. Seven years had passed since the hotel room tasting, and I decided to go find Cremisan and drink their wine again. I filled notebooks with questions and lists, forgetting about a line of Persian poetry that McGovern quotes: "Whoever seeks the origins of wine must be crazy."

With the *Wine Grapes* DNA research in mind, my wine pilgrimage became an expedition. I thought of Kermit Lynch's 1988 classic *Adventures on the Wine Route*, a lovely, quirky tour of France. Instead I would travel the original wine routes, from the Caucasus Mountains down to the Holy Land, across the Mediterranean and north through Europe. That's how winemaking spread throughout history. I'd interview scientists and archaeologists, seeking facts instead of just colorful myths. My father, may he rest in peace, worried when he felt I didn't have a plan in life. I had one now.

I was excited but nervous. I needed a wine science boot camp. Luckily José Vouillamoz agreed to let me spend some time with him in Switzerland, and there were other hopeful developments regarding Cremisan. A group of Israeli researchers had scoured the region for native grapes and performed DNA analysis on them. I wanted to learn what they found. A small distributor in the United States had also just started importing some Cremisan wines, but that wasn't enough to stop my trip.

In between booking flights I kept exploring the origins of wine, and learned that some modern alcohol laws have an ancient heritage. From underage drinking to DUIs and how to define drunkenness, societies have long struggled to balance wine with civilized behavior.

Two passages in the Babylonian Code of Hammurabi, the world's oldest law document, mention wine-related punishments for women. One concerned a priestess who had left a temple to drink (the punishment was death); the other may be the ancient version of holding barkeepers accountable for the size of a drink: "If a tavern-keeper (feminine) does not accept corn according to gross weight in payment of drink, but takes money, and the price of the drink is less than that of the corn, she shall be convicted and thrown into the water," reads one translation. So did male tavern-keepers get the same punishment, or did men just not do that job? No one knows.

Around 360 BC Plato described various commonsense prohibitions on drinking. "Let us then discourse a little more at length about intoxication, which is a very important subject, and will seriously task the discrimination of the legislator," said Plato's Athenian Stranger. "Shall we begin by enacting that boys shall not taste wine at all until they are eighteen years of age[?] [. . .] this is a precaution which has to be taken against the excitableness of youth;—afterwards they may taste wine in moderation up to the age of thirty."

The Stranger also suggested that "no one while he is on a campaign [i.e., war] should be allowed to taste wine at all . . . and that no magistrates should drink during their year of office, nor should pilots of vessels or judges while on duty taste wine at all, nor any one who is going to hold a consultation about any matter of importance; nor in the day-time at all, unless in consequence of exercise or as medicine; nor again at night, when any one, either man or woman, is minded to get children."

Throughout the third to fifth centuries AD, rabbis in Babylon debated how long it takes to walk off a bender. "A quarter of a lug of Italian wine inebriates a man; when a man is inebriated, he must not decide any legal questions; a walk neutralizes the effects of wine," reads the Mishnayot (a kind of religious common law). It goes on to suggest that if a Torah scholar is called on to nullify a vow, but has drunk too much, he must walk three miles (not just one) before making the decision. The Mishnayot also provides guidance on judging whether someone is tipsy or drunk. "What is meant by tipsy? If a man were compelled to speak to the king and still had sense enough to do so, he is merely tipsy; but one who would not be able to do this is considered intoxicated."

Ancient clay tablets from Mesopotamia also suggest the business of wine hasn't changed much in thousands of years. "If no good wine is available there for you to drink, send me word and I will have good wine sent to you to drink. Since your home town is far away, do write me whenever you need anything, and I will always give you what you need." That was written about 3,800 years ago in the now-dead language of Akkadian and found in Syria, in what was once the city of Mari, an important trading center on the Euphrates River, near the present-day Iraqi–Syrian border. Archaeologists have discovered more than twenty-five thousand tablets there, and many of them show how geography, politics, and money influenced the spread of wine. The ledgers from the Mari tablets kept a detailed record of who drank what:

"Expenditure of wine for dinner of Babylonians"

"Expenditure of wine for Babylonian generals when they received their gifts"

"Expenditure of wine as gift for Babylonian troops"

If you've ever struggled to budget enough wine for a large event, take heart. At least you didn't have to supply an army. An Egyptian king sent a firm reminder to Mari that wine was obligatory, not optional: "Be prepared for the arrival of the archers of the king, with much food at hand, wine, and everything needed. P.S. Be assured that the king is as well as the sun god in the sky; his soldiers and his [chariots] are in very, very good condition."

Wine was important enough that the Hittite army, which conquered cities such as Ninevah in biblical times, had an official position called *gal gestin*, or "chief of the wine stewards." And just like today, some people said the main purpose in life should be having fun. The tomb of a Hittite king who ruled around 1400 BC was decorated with a statue of him seeming to snap his fingers and the inscription "Eat, drink and be merry, for everything else is not worth that."

TASTINGS

Some early alcohol was made solely from grapes, but it could also be a kind of boozy goulash—a mixture of beer and wine. Peasants and royalty alike drank concoctions made from wheat, barley, rice, dates and other fruit, plus all sorts of spices. Over time wine, beer, and eventually distilled spirits settled into their own tribes. For thousands of years we have defined wine as something made entirely from grapes, so I didn't feel the need to explore the history of those other beverages. Maybe someday. Luckily you can get a sense from Dogfish Head Craft Brewery's line of Ancient Ales, created in close collaboration with Patrick McGovern:

Midas Touch: Based on analysis of what's believed to be the tomb of King Midas; "a sweet yet dry beer made with honey, barley malt, white muscat grapes and saffron."

Chateau Jiahu: Based on analysis of a Chinese tomb, made with "hawthorn fruit, sake rice, barley and honey."

Ta Henket: Made with ingredients listed in Egyptian hieroglyphics, a beer that contains an ancient variety of wheat, chamomile, palm fruit, and Middle Eastern herbs, and fermented with a native yeast strain from Cairo.

Kvasir: Based on a 3,500-year-old site in Denmark, and made with "wheat, lingonberries, cranberries, myrica gale, yarrow, honey and birch syrup."

3

Cremisan

Therefore God give thee of the dew of heaven,
and the fatness of the earth, and plenty of corn and wine.
—Genesis 27:28 (KJV)

I was nervous walking around one of Jerusalem's old neighborhoods on a clear, fresh April morning, but not because of anything around me. The shops were upscale, the restaurants had tempting menus, and the falafel and shawarma were tasty at the corner joint. The jitters came from the rapid approach of my vineyard reckoning. Seven years after tasting that Cremisan wine in the hotel room, I had a visit scheduled. My thoughts bounced between apprehension and excitement. Is it even possible to remember a taste from so long ago, or was my passion based on faulty memory? I pondered other scenarios. Maybe the wine used to be glorious, and now it was mediocre. Then again, it may have gotten better. Some Israeli newspaper stories explained that Cremisan had brought in a famous Italian winemaker to help.

Another question was bothering me, too: Could any of the old wine texts I was reading help me understand what I had tasted, and what the ancients drank? Or was it all seductively poetic speculation? A colleague suggested I talk to Aren Maeir. He is a renowned archaeologist who has led the dig at the Philistine city of Gath—reputedly the home of Goliath—for more than two decades. "Gath" can mean winepress in Hebrew, and the region was a major center for wine production in ancient times. Maeir's team has found whole sets of wine-related vessels at the site, from large storage containers to delicate clay serving flasks with inscriptions. Just as the French take morning coffee, the Philistines sipped wine from bowls. In my correspondence with him, I noticed that Maeir ends his emails with a quote from philosopher Karl Popper that seemed appropriate for my wine investigations: "Whenever a theory appears to you as the only possible one, take this as a sign that you have neither understood the theory nor the problem which it was intended to solve."

Maeir suggested meeting at a cafe on Emek Refaim Street, a trendy neighborhood not far from the Old City walls. He's a slightly stocky man in his fifties, of average height, with very short graying hair and an easy sense of humor, which scientists sometimes lack. I asked if ancient texts are still relevant today. "Just in general, when you think about the ancient world, you have to realize that these people are exactly like us, save for that they don't have the technology that we have," Maeir said. "The same emotions, the same needs, the same everything."

I mentioned the claim, which I had read in *The Oxford Companion to Wine* and elsewhere, that wine essentially vanished from the Holy Land after the Muslim conquest, except during the Crusades. He said there was no question winemaking continued, though on a smaller scale than before. "You had Christians and Jews living in the land. You can't expect that things disappeared,"

Maeir said. With a quick smile he added that the Bible forbids adultery and Prohibition banned alcohol. Did Christians stop sinning, or Americans stop drinking? Maeir told me that not all Muslims, particularly in the early centuries after the Arab conquest, abstained from wine. Under Islamic rule, Christians and Jews were also allowed to make, sell, and consume wine if they paid a special tax. In other words, not everyone observed all the rules about wine, or probably anything else.

Henry Maundrell, an English clergyman and Oxford University scholar, wrote a book titled *A Journey from Aleppo to Jerusalem at Easter A.D. 1697* that perfectly captures the contradictions of life in the region at that time. He found one coastal tribe that changed its religion to match every business opportunity. "With Christians they profess themselves Christians; With Turks they are good Mussulmans; With Jews they pass for Jews . . . All that is certain concerning them is, that they make very much and good Wine, and are great Drinkers." Maundrell also wrote of visiting a monastery in the region, observing, "It is a place of very mean Structure, and contains nothing in it extraordinary, but only the Wine made here, which is indeed most excellent."

At the end of our conversation, I thanked Maeir for his time, and he offered to give me a tour of Gath on a future trip. Later I casually looked up Emek Refaim Street to see what it meant, and laughed. It's named after the biblical Refaim Valley, reputedly home to a race of giants, and is where King David won battles over the Philistines. I had listened to an archaeologist tell Goliath stories at a cafe on Valley of Giants Street, and of course there were the parallels between the story of David and Cremisan, the tiny winery going up against far bigger competitors.

Next I visited Uri Mayer-Chissick, a historian of Jewish and Arabic food and culture. He lives with his wife and kids on a kibbutz, or communal farm, near the Sea of Galilee in northern Israel. He guides foraging tours, gives talks on healthy eating, and sells a selection of locally produced spices, oils, teas, and organic produce. He was stirring an enormous pot of chickpeas in a small kitchen when I arrived.

Jews, Christians, and Muslims share many of the same foods, Mayer-Chissick told me. His PhD thesis was on traditional medicine in the medieval Arab era, and wine production, too. He said the *Oxford Companion* was wrong to claim that winemaking vanished from the region for centuries. "[Muslims] still drank wine. The rich people drank wine, although it was forbidden," he said. One Ottoman ruler was known as Selim the Sot for such habits, and records from the time of the Crusades up to the nineteenth century show that many Jews and Christians sold wine, though it was often made at home.

Mayer-Chissick read passages from his research to me. In 1384 a Christian source remarked, "In Gaza each person has his own wine." In 1488 a Jewish source observed that "in Jerusalem people drink the living wine," meaning it was less than one year old and wasn't diluted with water. In 1818 a Jewish man visited the northern Israeli city of Safed, which dates to biblical times, and he described five kinds of wine, including some for the rich that was fifteen to twenty years old. A Christian pilgrim to Jerusalem in the seventeenth century found "a wine diluted with snow," which had probably been brought over from the Lebanese mountains. Mayer-Chissick told me Jewish scholars discussed many aspects of buying, selling, and producing wine, saying, "If you buy wine from somebody, if it's rotten or made to vinegar, who needs to pay? There are many stories about that. We know that they had a living wine—a wine that was very, very young—and a wine that was very old. There were three categories: new wine, the youngest; old wine, that's between one and three years; and the oldest wine was three years and more."

Mayer-Chissick doesn't follow modern wine trends, but he had some general thoughts on why the *Oxford Companion* claimed wine vanished from the area after Mohammed's victories in AD 636. Social and academic exchanges between Muslims and Jews or Christians in the region have dwindled in recent decades, he said, and few people pay attention to a more complex past. He said Jews and Christians were certainly discriminated against under Muslim rule, yet some were also trusted advisors or doctors to the ruling class. When the Crusaders invaded, Jews allied themselves with Muslims, fearing that Christian persecution would be even worse.

In the years after Mohammed and the rise of Islam, there's no doubt that many of the vast Egyptian vineyards from earlier eras turned to producing table grapes, but winemaking never vanished. Despite all the war and persecution, a never-ending stream of both Jewish and Christian pilgrims kept returning to the Holy Land, and some stayed. An 1867 book by William Wyndham Malet specifically praised wine made near Cremisan. "There is red wine from Cyprus, but the Pilgrim prefers the white wine of Bethlehem, also on the table; this is a good 'dry' wine."

I said good-bye to Mayer-Chissick and left feeling somewhat encouraged. Cremisan really could be using ancient, native Middle Eastern grapes.

The day arrived. I stood on a Jerusalem street corner with

David Silverman. He is an excellent Israeli photographer and wine lover who covers the industry, and I thought his experience would help me gauge the Cremisan wines. Soon Amer Kardosh pulled up in a modest car. Kardosh is one of Cremisan's regional distributors and he had driven from his home in Nazareth to give us the tour. He told us a little about himself as we drove. A friendly, energetic man in his late forties, he is part of the dwindling community of Middle Eastern Christians.

After escaping traffic we came to a narrow, curving road that followed a ridge between Jerusalem and Bethlehem, about three thousand feet above sea level. Terraces of olive trees stretched far down the mountain, like a twisted, rocky ladder. Groves of poplars, pines, and cedars covered the higher elevations, and the valley had a timeless feel, as if a hermit might casually emerge from a cave at any time. We stopped at a simple gate. "This is Cremisan," Kardosh said, and I squeezed the seat and almost jumped out the door towards the limestone buildings. By one measure the trip from downtown Jerusalem took only fifteen minutes; by another it had taken years.

We walked past a four-story monastery building without see-ing or hearing anyone. It was about a hundred feet long, and made of limestone blocks covered with sun-faded stucco. Classic arched windows and the size of the monastery suggested that it was de-signed to house dozens of monks. At the edge of the walkway a glorious panorama unfolded across the valley, stretching down five hundred feet or more. Vineyards first, then I counted two dozen lines of densely packed, curving rock terraces, each about a quarter-mile long, with scores of old olive trees on each level. They held miles of hand-laid stones. "Very, very, old terraces. No cement, just rocks," Kardosh said. Cremisan's side of the valley was one of the few undeveloped areas near Bethlehem or Jerusalem. On the other side, a mile away, new settlements and apartments clogged the hill. "So what was Cremisan's wine like in the beginning?" I asked.

Kardosh said Cremisan was built by the Salesians, a Roman Catholic order founded in the mid-1800s to help poor children. Italian monks first came to the area in the 1860s and bought land, using the natural limestone caves around the ruins of a seventh-century church to store wine and other agricultural goods. The Salesians contributed money and people to build the monastery in 1881, and at first made sweet wine to celebrate the Mass. Then other churches and monasteries asked for that wine, and dry wine, too. "They started to sell it to other monasteries. Later it became a small factory." The business grew to support an orphanage, a technical school, another monastery, an agricultural school, a kindergarten, and a bakery. "So all the money that we earn goes back to the community in different ways," he said. The winery prospered in the 1980s and 1990s, selling to Christians, Jews, and some Muslims while rising to a peak production of more than six hundred thousand bottles per year. The Cremisan gift shop was often packed on Saturday mornings.

We followed a path towards the first terraces, past a small organic vegetable garden that helps feed the monks. Cremisan nuns run a school in a separate building, too. Turning a corner, an enormous, gnarled grapevine the size of a small tree sprawled up and out from its plot next to some stone buildings. "Look at this!" I said. It was about eight feet tall and two feet around at the base, with a canopy of twisted limbs extending out fifteen feet. Green leaves were just beginning to emerge.

"This is an old vine. Unbelievable. It's still growing, still giving us grapes," Kardosh said, adding that it is reputed to be more than one hundred years old.

We walked along the ridge road towards the main vineyard and passed a small grove of almond and plum trees. I asked whether the monks had brought grapes with them from Italy in the late 1800s, or if they used the local grapes. "For sure both," Kardosh replied. But for decades Cremisan did not focus on the local varieties, using them only to blend with wine made from European grapes.

Trying to avoid sounding rude, I asked the question that had puzzled me since tasting Cremisan in the Amman hotel room— why do so few people know about their wines? Kardosh sighed. The Salesians ran it in an almost otherworldly fashion. For decades they didn't export or do much marketing to anyone beyond Christians in Israel, the Palestinian Territories, and nearby Jordan, which has a large Palestinian population. Time was an issue, too. "Here, to do something, you need . . . years," Kardosh said, since the monks painstakingly debate every issue. He added that helping orphans is the most important mission for Salesians, and then explained his own ties to Cremisan—his father was an orphan who grew up to become one of Cremisan's wine distributors. Kardosh took over that position in 2001, after his father died, even though his degree is in electronics. "I've known them thirty, forty years," he said of the monks.

I realized my romantic vision of Cremisan's vineyards had missed a crucial point. I obsessed about how the wine tasted, exactly what grapes they were using, and how those grapes related to the history of wine. Perfectly normal questions from a twenty-first-century wine nerd. But Cremisan had other priorities. They made wine as a part of rituals that go back to the earliest days of Christianity. To devout Christians (and Jews), grapes are one of God's gifts to men and women, and can be symbols of ecstasy, temptation, and loss. The Bible mentions wine, grapes, or vines hundreds of times, such as this poignant passage in Isaiah called "The Song of the Vineyard."

> My loved one had a vineyard
> on a fertile hillside.
> He dug it up and cleared it of stones
> and planted it with the choicest vines.
> He built a watchtower in it
> and cut out a winepress as well.
> Then he looked for a crop of good grapes,
> but it yielded only bad fruit. [. . .]
> When I looked for good grapes,
> why did it yield only bad?

There's also this famously dark vision of retribution and redemption from Revelation, which inspired "The Battle Hymn of the Republic" and Steinbeck's *The Grapes of Wrath*: "So the angel thrust his sickle into the earth and gathered the vine of the earth, and threw it into the great winepress of the wrath of God. And the winepress was trampled outside the city, and blood came out of the winepress, up to the horses' bridles, for one thousand six hundred furlongs."

The monks didn't care what wine critics or magazines thought.

The vineyards and olive groves helped the poor and provided wine to celebrate the Mass. They didn't need to do PR. Why would a score from a critic matter on Judgment Day?

The second Palestinian Intifada, or uprising, was taking place when Kardosh started working as the distributor. "When there is [an] uprising, everything is stopped totally," he said. Cremisan had trouble transporting wine, and people who had been visiting the monastery gift shop for decades no longer came. Israeli authorities started building a security barrier in the area, and the dispute continues over how it cuts through Cremisan lands. The case has gone all the way to Israel's Supreme Court. Kardosh said that if the wall is fully built, Cremisan workers who now have a five-minute walk from a nearby village will have a much longer commute by car, creating numerous problems.

On top of everything else Cremisan has had to juggle the religious, cultural, and legal sensitivities of suppliers and customers. The word Byzantine originated in this part of the world, and such entanglements persisted long after that empire fell. Some of Cremisan's buildings are technically inside Jerusalem's district—and under Israeli control—and others are in the Palestinian West Bank. Most of the grapes for the wine come from Palestinian areas, but some vineyards are in Israel. That created a bureaucratic dilemma over how to label the wine for export. Some European nations are attracted to the wine's Judeo-Christian links, but in Japan it's marketed as a Palestinian product. Other places forbid using the word Palestine, since technically there isn't such a state. One rabbi who liked the wine suggested they would sell far more if it were labeled kosher; Cremisan politely demurred, since under kosher laws the monks wouldn't be allowed to touch the winemaking tanks.

As outside pressures mounted, so did inward ones. By 2005 the winery was faltering. Sales had fallen by more than 50 percent. Some vintages met the established quality standards; others didn't. To survive Cremisan needed to change, and seek new markets. That led to a partnership with Italian wine consultant Riccardo Cotarella and Civielle, an organic vineyard cooperative located between Venice and Milan. The Italians began to help Cremisan understand and appreciate what an extraordinary resource they had in the local grapes. "We had used the grapes before but we didn't understand their significance," Kardosh said.

Several grape varieties were sent to Italy for DNA analysis, including Dabouki, Baladi, Hamdani, and Jandali. The tests found a genealogy different from any well-known Western grapes, though perhaps related to some in Spain. That mix could be because seafaring people carried Middle Eastern vines across the Mediterranean thousands of years ago. The vines bred with local varieties, leaving traces of their DNA. "We have something here. We have grapes that nobody has," Kardosh told me. The monks made a renewed commitment to the winery and to supporting the many families that have worked in the vineyards for generations.

Kardosh, Silverman, and I continued walking down the ridge road, past vineyards on our right, a soccer field and basketball court for local youth on the left, and more groves of cedars and poplars on either side. Kardosh smiled. This was a secret place for lovers to come and talk, away from family and friends, he said. We came to a hundred-yard-long terrace of Baladi vines. The first green leaves were just bursting out, and Kardosh introduced me to one of Cremisan's employees, forty-year-old Baha Darras. Everyone in Darras's family—including his father, grandfather, and grandmother—had worked in the vineyards at times, but none drank wine since they're Muslims, he said. Another longtime Muslim worker told me he can judge the quality of a vintage by color and smell. "We have Christians and Muslims working together," Kardosh told us proudly.

Workers were hand-weeding the vineyard. Ladybugs crawled along the leaves. They're a natural version of pest control since they eat aphids. I pointed to the bugs but Kardosh didn't understand the name I used. Here everyone calls them Moses bugs, or Moses's cow. The venerable Jewish–American newspaper the *Forward* untangled the linguistic trail, noting that the "Lady" part of our term refers to the Virgin Mary in many European languages. In French the bugs are sometimes called *la vache de la Vierge* ("the cow of the Virgin"), while the German is Marienkäfer ("Mary's beetle"). The *Forward* suggested that Jews were more comfortable naming the humble bug after Moses.

We walked a little farther and arrived at the winery. It had moved from the monastery basement as business expanded in the 1970s. The new building is perched on the edge of the ridge, next to a vineyard, with old cast-iron wine-pressing machines casually displayed in various spots. We went inside to an enormous room with twenty-foot ceilings. The Italians had donated a new bottling machine, but a large nineteenth-century copper brandy still sat in one corner, too. Stone arches framed the entrance to the fermentation room, which had old cement-lined vats sitting next to stainless steel ones. Dozens of five-foot-tall oak barrels were packed under limestone arches lining one wall, next to some less than half that size.

Kardosh introduced us to Laith Kokaly, Cremisan's new twenty-nine-year-old winemaker, and I realized for the first time that monks no longer make the wine. In recent years there were fewer and fewer young candidates to replace those who got sick or died, and Cremisan now had only seven active monks. Kokaly, a Palestinian Christian from the nearby community of Beit Jala, studied winemaking in Italy for three years. Cremisan paid for the training. Kokaly told me his grandfather made wine from local grapes many years ago, and Kardosh nodded. In the past almost every Christian family in the area made their own wine. Kokaly poured us some of the newest Dabouki white wine, a 2014. Crisp and slightly sweet, it had hints of citrus and caramel.

"I love the aroma," Silverman said, and we both thought it was a refreshing, summertime type of wine. Next we tried Cremisan's blend of Hamdani and Jandali white grapes, which was crisp but minerally, and with far more depth and flavor than the Dabouki. It was a lovely, balanced wine, one that Silverman thought was different from the typical white wines being made in Israel, which were using European grapes.

Kokaly told me that with Cotarella's help Cremisan improved all aspects of the winemaking, from the vineyards to fermentation and aging. The response to the wines made with native grapes was positive enough that Cremisan stopped using many European varieties. "Enough. Enough of Chardonnay," Kardosh said. I loved the Cremisan whites, but could hardly wait to taste the red wine that had captivated me years ago. Kokaly poured us some new 2014 Baladi from tanks, and Silverman and I sniffed and tasted.

I was almost speechless. The red was a decent dry wine, but unremarkable, with none of the spiciness and depth that had first attracted me. I felt a little unsteady, and it wasn't just the alcohol. My memories had crashed into reality. We tried some of the bottled Baladi. I had the same reaction. It was an OK wine, but not the kind anyone obsesses about for years.

I asked a few more questions about the 2008 reds, but didn't push because of a fundamental problem: Kokaly wasn't that wine-maker. Cremisan was focused on the future, not the past. In retrospect I had missed a warning sign. When the sommelier for the restaurant Ottolenghi praised Cremisan on the *Wine Grapes* web-site, it was for the white wine. Gal Zohar had called the Hamdani-Jandali blend "crisp, fresh and with a nice complexity to it . . . very original and dangerously drinkable, too." He didn't mention Cremisan's red.

We walked back to the old monastery and stopped at the gift shop, but it was locked. Someone went looking for the key. I saw a narrow garden path near the gate and headed down it, more out of idle curiosity than anything else. At the end was an old hand-powered cast-iron winepress—I guessed from the early 1900s—and in a little clearing to the left, under some tall poplars, were ancient limestone grape- and olive-pressing troughs, stone rollers, and wine-pressing troughs, hundreds or perhaps thousands of years old. I sat in the garden, dumbfounded. A French winery would die for such a museum exhibit; this one didn't even have plaques or any explanations.

I thought of something Aren Maeir had said. Whether you think the Bible and Torah are about specific men and women who once lived or not, there is no question that those books were writ-ten by real people who walked through valleys just like the one I was looking out over, and who drank wine in these very hills thou-sands of years ago. I was looking at some of the evidence.

I stayed in the garden for a few minutes, daydreaming of the people who used those stone tools, then headed back to the gate. The gift shop key was found and I bought some wine and locally produced almonds. When I asked Kardosh about the garden exhibit, he casually told me that all of it had been found on the Cremisan property, or in the case of the cast-iron press, used by the monks in the past. But no one knew much about the exact dates of the artifacts. They had too much else to worry about.

I left Cremisan exhilarated, puzzled, and feeling like I was missing something. The wine I drank in 2008 remained a mystery. It was probably made in 2006—just before the Italian consultants got involved, so I'd tasted one of the last vintages made by the monks. Now that bottle seemed as unreachable as wine made two thousand years ago. No one knew exactly what was in it.

I thanked Kardosh for the tour, said good-bye to Silverman, and put the Cremisan questions on hold, wondering what my next meeting with an Israeli scientist would reveal. I was in for more surprises.

TASTINGS

Listed below are some of Cremisan's wines available in the United States. The wines are produced organically, without pesticides. All are about twenty dollars. These are fresh vintages meant to be drunk right away. It's not clear yet how they age.

Star of Bethlehem Hamdani & Jandali blend (white)

Star of Bethlehem Dabouki (white)

Star of Bethlehem Baladi (red)

67 Wines & Spirits sells them at their Manhattan store located at 179 Columbus Ave., New York, NY, 10023, and online at www.67wine.com.

Many Israeli and Middle Eastern restaurants on the East Coast sell Cremisan wines, too, such as Tanoreen, a wonderful restaurant in Brooklyn.

In addition to red and white wines, Cremisan makes a brandy and an altar wine that are available in parts of the Middle East. See terrasanctatrading.com for more details.

4

Israel's Forgotten Grapes

I found Israel like grapes in the wilderness.

—HOSEA 9:10 (KJV)

The Israeli city of Ariel sits in the heart of the Palestinian West Bank, which makes it a political flash point. News stories often refer to it as a settlement, so part of me expected some tiny little outpost. Instead I found a rapidly growing university and community of twenty thousand people. Shivi Drori is one of them. He's a scientist and winemaker doing research on indigenous grapes.

When I first read about Drori's work it seemed like a modest effort. The Ariel winery fits inside a metal shipping container, with test batches fermenting in metal kegs. But the clean, well-equipped lab was filled with diligent young scientists. I revised my opinion—this was a serious project.

I told Drori that I'd visited Cremisan the day before. "Nice," he said in a tone both businesslike and enthusiastic. He's like that, a crack scientist and former paratrooper, friendly but sure of himself. I asked how he got interested in native grapes, given that nobody in the modern Israeli wine business used them. Cremisan—located in the Palestinian West Bank—was the only such winery. The biblical stories of vast ancient vineyards had intrigued Drori, even though many people insisted the local grapes of the present were fit only for eating or for juice.

"This is a very old dream of mine. I'm a winemaker, eleven, twelve years. And from the start I thought about doing let's say, special things. Not just the Cabernet, Merlot, whatever."

I started to ask a question. Drori gently but firmly asked me to wait until he finished the background story. Starting in 2011, with funding from the Jewish National Fund, Drori's team analyzed all the local grapes that had names, even the table grapes. "And we found out a few of them could actually be used for wine production. The whites are very interesting," he said.

Drori's team did DNA analysis on the grapes, cataloged archaeological sites with grape remains, analyzed sugar content and acidity (key benchmarks for wine), created sample micro-vintages, and did rigorous taste tests to identify the flavor profile of each variety.

Such comprehensive research into the origins of obscure grapes rarely happens. Most grape science around the world focuses on improving or increasing production, or controlling diseases and pests. In other words, things that help wineries make money.

The project also did historical research on grape names, and catalogued leaf structure and grape seed shapes. "We have some nice findings about the historic origins of the Jandali and Hamdani dating to the fourth century. They were called Godali and Haldali," he said. Drori's research shows that there's evidence they were used to make wine for centuries. In about 1600 Rabbi Menahem de Lonzano wrote that "up to this day there are two sorts of wine in Jerusalem: Jandali and Hamdani wines," and likened their tastes to different women's personalities.

Drori's team identified nineteen local varieties with wine grape qualities. The project expanded into a nationwide search for wild grapes, with populations pinpointed by GPS. "Meaning we are walking all through Israel, from the north to the south, collecting every vine we can put our hands on. Out of these, we have today an additional one hundred varieties we found that are unique to Israel," he said.

He showed me some papers and slides on his laptop, and explained that DNA analysis found some links between the Holy Land grapes and those in the Caucasus. That could represent a branch of the grape family tree, which grew as winemaking spread out of the mountains and down into the Holy Land. Drori said he was collaborating with other scientists on more research.

I mentioned that I'd be visiting the Caucasus, where most experts believe wine grapes were first domesticated. "It's a theory which I hope to demolish soon," Drori said dismissively. "We believe that Israel was a wine domestication point." That's a common dream among grape scientists. Everybody wants their country to be first. Scientists in a half-dozen countries have tried to disprove the Caucasus origins theory, with no real success. New evidence is always possible, however, and the goal isn't so far-fetched. In 1997 German researchers did the same thing for einkorn wheat by comparing the DNA from scores of wild and domesticated strains. They traced all modern domestic einkorn to a single wild population in southeastern Turkey. Drori was trying to find a grapevine version of that.

He had some practical insights, too: "I believe that Israel will finally understand that it is a hot region. Here winemakers can either fight the sun and try to produce European wines, or go with the sun." In other words, they could keep planting cool-climate European grapes, or use native grapes that evolved for a Mediterranean climate. Drori also thinks Cremisan's current efforts, and even the tests from another Israeli winery, are misguided. They use native varieties but only to produce European style wines. "That doesn't make a lot of sense," Drori said.

We left the lab and went over to the basement of another building, where precious samples of grape seeds found at archaeological sites are kept. A young woman in a white lab coat opened a cabinet, pulled out a small oblong box, and tipped a single dark black seed into her gloved hand. The seeds were from the era of King David. The thought that someone had picked those grapes about three thousand years ago sent a chill up my spine. With more work—and some luck—Drori hopes to match the DNA of an ancient grape seed to an existing native variety.

"At the end of this year, we will have some notion about which grapes were used for wine production here in ancient times. We have remains from the Temple Mount, and we have them from Ashkelon, from the north, all over Israel," he said. An Israeli winery is working on a Hamdani and Jandali blend, too, so it looks like Cremisan will soon have competition.

The visits to Cremisan and Drori exceeded my expectations, and confounded them. Clearly the *Oxford Companion* was wrong to state that Israel had no native wine grapes. I'd worshipped the memory of a Cremisan red wine, but now I liked their whites best. And I saw Drori's point: the Cremisan wines were made in a

European style, aged in oak or stainless steel. I was back to wondering what ancient wine really tasted like.

Vouillamoz and Drori are first-rate scientists, but with a mix of emotions and new questions I sought out viticulture second opinions, and called Sean Myles, the lead author on a widely cited 2011 paper on grape biodiversity published in the *Proceedings of the National Academy of Sciences*, a top journal. He has a Masters of Science degree from Oxford University and a PhD in genetics from the Max Planck Institute for Evolutionary Anthropology in Leipzig. A native of Canada, Myles now teaches at Dalhousie University in Nova Scotia. First I asked if there was any justification for planting just a few grape varieties all over the world. He reeled off a botanical sermon about rampant viticultural apartheid.

"If applied to any other category you'd say this is just plain old racism. A little bit of wild ancestry? Ah, you're still a hybrid. You're inferior to the noble European grapes," Myles said, and then provided an example of the industry mind-set. He had met with leaders of a large American wine company, which sells billions of dollars worth of wine per year. Sitting around a table with the heads of the company, Myles remarked that about twenty grape varieties take up virtually the entire world wine market. One of the executives replied that the number was actually six. Myles eventually shifted his research priorities from grapes to apples because resistance to change in the wine industry was so great. He and his wife recently opened a cider bar in Nova Scotia.

I called Andy Walker, the Louise Rossi Endowed Chair in Viticulture at the University of California, Davis, one of the best schools in the world for wine grape research. He was mild-mannered but equally firm. "We're still caught in that trap of saying, well, there are only ten good varieties in the whole world, and that's it. Anyone who's drunk wine around the world realizes this is a complete fallacy," Walker told me. "There are wonderful wines to be made everywhere from a huge number of varieties. But it's a marketing scam that we ended up with ten varieties that are [supposedly] destined to be the best in the world." All good wine grapes match a particular environmental niche, Walker said.

Independently of one another, some of the best grape scientists in the world agree that there is no reason to make wine from only a few varieties of grapes. As UC Davis scientist Carole Meredith observed about wine after she left academia, "Can there be any other business where there's so much bullshit?" It all made me think. Even beginning gardeners know that orange trees don't do well in Vermont and apples don't thrive in south Florida. Yet thousands of vineyards around the world have rammed the famous French varieties into an equally unsuitable range of habitats. With no disrespect to the best winemakers, the idea that a few French varieties are "noble" has plenty of flaws. The earliest mention of the term in relation to grapes is in the Bible: "Yet I had planted thee a noble vine, wholly a right seed: how then art thou turned into the degenerate plant of a strange vine unto me?" (Jeremiah 2:21, KJV) and other passages suggest particular valleys had especially choice purplish grapes. In the seventeenth century Shakespeare uses the term in *All's Well That Ends Well* as a pun, when the old nobleman Lafew tells the king, "Oh, will you eat no grapes, my royal fox? Yes, but you will, my noble grapes."

The French eventually used "noble" to refer to six grape varieties (Cabernet Sauvignon, Chardonnay, Merlot, Pinot Noir, Riesling, and Sauvignon Blanc) and some others were deemed "common." It all had about as much factual basis as the idea that kings and queens were better than other mortals, or the mistaken belief once common among winemakers that a red grape couldn't produce white offspring. Over time wine lovers (and wine sellers) simply became attached to the idea of noble grapes.

Given all the centuries of hype, I wanted one more scientific perspective about grapes, even after talking to Myles and Walker. Whether you're writing about finches, tortoises, earthworms, or any other creature, it always makes sense to see what Charles Darwin had to say on the subject. It turns out that he was fascinated by grapes, too. Though grapevines are often presented to wine lovers as a kind of viticultural pinup—the fruit arranged in perfect, bountiful rows—the great British scientist found a curiously complex plant.

To better understand its origins, Darwin placed a bell glass over a young Muscat grape in a hothouse and watched the shoot move every day. The plant was fidgety, and he found that odd. "Had it not made at least three revolutions whilst the sky was uniformly overcast, I should have attributed this slight degree of movement to the varying action of the light," Darwin wrote in an essay titled "On the Movements and Habits of Climbing Plants." He continued, "It has often been vaguely asserted that plants are distinguished from animals by not having the power of movement. It should rather be said that plants acquire and display this power only when it is of some advantage to them . . ."

Grapes are lianas, a category of tree-climbing vines. Kudzu and rattan—both of which can grow hundreds of feet long—are lianas, too, and suggest how grapevines might behave if left to their own devices. Their ancient relatives were able to climb tremendous heights, reaching the places with the most sunlight, but as Darwin wrote, they didn't have to spend very much energy doing so: "Plants become climbers, in order, it may be presumed, to reach the light, and to expose a large surface of their leaves to its action and to that of the free air. This is effected by climbers with wonderfully little expenditure of organized matter, in comparison with trees, which have to support a load of heavy branches by a massive trunk."

Grapes use their wispy tendrils almost like rock climbers use their hands. "The tendril strikes some object, and quickly curls round and firmly grasps it. In the course of some hours it contracts into a spire, dragging up the stem, and forming an excellent spring. All movements now cease. By growth the tissues soon become wonderfully strong and durable. The tendril has done its work, and has done it in an admirable manner," Darwin explained, and thus the vine is able to "easily pass from branch to branch, and securely ramble over a wide, sun-lit surface."

Some botanists even compare lianas to a parasite, since they compete with tree leaves for the sun while their roots also take water from below. In grapes this results in more energy to produce large bunches of fruit, which attract birds and other small creatures, who subsequently spread the seeds. As if all that isn't enough, tendrils can evolve into either a climbing tool or flowers and grapes. Darwin remarked that even the boldest believer in evolution "would never have surmised that the same individual plant, at the same period of growth, would have yielded every possible gradation between ordinary flower-stalks for the support of the flowers and fruit, and tendrils used exclusively for climbing. But the vine clearly gives us such a case." In other words, various limbs on the same plant multitask in different ways. That might not seem like a big deal, but consider the human parallel: hands that could grasp tools *or* be a reproductive organ. Darwin's essay included two drawings to illustrate the point. The one on the left shows tendrils B and C growing from the grape stalk. In the illustration on the right tendril C bursts into flowers, while B attaches to another branch to help support others heavy with grapes. Amazing.

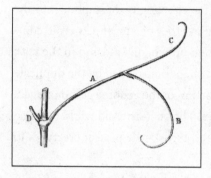 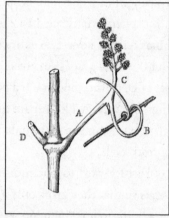

To Darwin, the multitasking tendrils told a profound story of grape evolution because they illustrate a moment of transition between primitive plants and later, more complex varieties. The earliest plants didn't even have flowers or seeds—think lichen and moss. Fast-forward more than a hundred million years and there are flowers and seeds everywhere.

So how did grapes become so adaptable? Dodging mass extinctions may be part of the answer. The earliest lianas survived the K–T mass extinction event of approximately sixty-six million years ago, which helped kill off large dinosaurs as well as many plants and sea creatures. A massive asteroid traveling at more than forty-four-thousand miles per hour hit the Yucatán Peninsula, creating a crater that's estimated to have been about one hundred miles wide. The impact was at least a billion times more powerful than the atomic bomb dropped on Hiroshima, and it triggered tsunamis, megastorms, and a worldwide cloud of ash and debris. That was only part of the bad weather news of the K–T extinction era.

Before, during, and after the blast an enormous group of volcanoes in India were erupting, spewing toxic sulphur into the atmosphere. Many scientists believe the volcano and asteroid combination was a one-two extinction punch, on land and in the seas. But the early grapes survived it all, perhaps aided by the duplicate set of DNA they possess. If a portion of the genome gets damaged, there's a backup, like a spare tire. The post-asteroid world was also filled with smaller mammals and birds—the perfect creatures to spread seeds.

Fast-forward to our modern vineyards. We prune and train vines so that they grow only a few feet tall, and we provide stakes or a trellis to hold them up. We eliminate other plants or trees that would compete for water or sun. The result? A plant that evolved to survive global extinctions and grow dozens of feet high now devotes all of its energy to just two things: putting down deep roots (thus guaranteeing a steady supply of nutrients and water), and producing tasty fruit. By removing all other distractions from the lives of modern wine grapes, we turned them into prodigious flavor factories. The wild grapevine was an absolute free spirit, able to grow this way and that, spread its seeds far and wide, and have sex with other vines it rambled across. We've short-circuited all that.

For thousands of years cultures have devoted enormous amounts of energy to grape health and well-being. Just like our ancestors before us, we carry them to new fields, fertilize and water the vines, and compose worshipful poems and prose about them. So it seems fair to ask whether grapes are our masters, and not the other way around.

5

The Wine Scientist

Mark the sun's heat; how that to wine doth change,
Mix'd with the moisture filter'd through the vine.
—DANTE ALIGHIERI, *THE DIVINE COMEDY*, AD 1320

I flew to Switzerland to meet José Vouillamoz in the early spring. A two-hour train ride into the Alps wound along the Rhone River, passing a surprising number of vineyards, a waterfall that burst from a rocky mountainside, and field after field of rapeseed, cadmium yellow against the backdrop of snowy peaks. The Swiss also harvest large quantities of dandelion greens for salads and cooking, and while there I learned the origin of the name. It comes from the French *dents de lion*, or "lion's teeth," after the spiky green leaves around the flowery head.

I mused over what Vouillamoz might have to say about the wine grape family tree, and considered my own genealogy. My mother's ancestors came to rural Tennessee and South Carolina from England and France in the 1700s. My southern kinfolk have reams of records—old Bibles, letters, photographs, marriage certificates, details of who fought in the Revolutionary War, the Civil

War, and so on. They love to talk about all of it. My father's side of the family came from Ireland and Poland. Records for them are virtually nonexistent. We know a little about my Irish grandmother, and nothing for certain about my Polish grandfather—not even his family name in the old country, or what city he grew up in. That set of relatives doesn't talk much about the past. Scientists have long faced similar problems with grapes, but DNA research is revealing stories about long-hidden vineyard ancestors.

The train arrived exactly on time in Sion, the small city on the Rhone where Vouillamoz lives. The Italian border is thirty miles away. We met for lunch at one of his favorite restaurants, Le Coq en Pâte. He ordered a beautiful 2009 Syrah from the local winemaker Simon Maye & Fils, full of fruit and white pepper flavors. We ate a carpaccio pâté of eggplant, sliced translucently thin. Served with olive oil, it was a luxurious vegetable version of prosciutto.

Vouillamoz is a slight man with short dark hair, glasses, and a quick sense of humor. Some people see him as a kind of John James Audubon for endangered grapes, traveling the world to discover and preserve old varieties, and alerting wine lovers to the news. Whereas McGovern focuses on exploring archaeological sites and recreating ancient beverages, Vouillamoz performs DNA analysis, seeks vines that are alive but endangered or forgotten, and then tries to revive them. He is also a rebel who enjoys listening to (and supporting) alternative and underground music. One blogger on Facebook includes a long list of rants against music industry forces, alongside the line "I'd like to thank José Vouillamoz from Switzerland who single handily [sic] paid for me to fly to Sydney and back to record the Mars Volta back in 2011."

As we ate Vouillamoz told me the latest news, about resurrecting the forgotten Grosse Arvine grape with a local winemaker. "He was going in old vineyards trying to identify old vines. And he marked sixty vines. I went with him, with another viticulture expert," Vouillamoz said. Some of the vines were diseased, so they selected the best twelve. "Out of these twelve he planted a new vineyard, five hundred square meters. And a few weeks ago he invited me for the tasting of the first wine. It was honestly amazing. Honestly amazing. Acidic; maybe more for serious wine lovers, Riesling type of guys. But the quality was . . . wow! We were professionals saying what an amazing job he'd done with this forgotten variety. The last time anyone made a wine from it was one hundred years ago."

Vouillamoz doesn't own a car because he feels they are polluting and unnecessary, given the country's excellent public transportation. He dresses elegantly in dark jackets or suits with black or white shirts; has three teenage children, two boys and a girl; and is endearingly modest, especially given the avalanche of praise *Wine Grapes* has received from critics all over the world. *Decanter* called it "a magnificent achievement: colossally informative, illuminating and intriguing." The *Wall Street Journal* wrote, "This astounding work of scholarship . . . will advance anybody's wine education by huge, passionately nerdy leaps," and the *New York Times* described it as a must-have, despite the $175 list price. "A masterly work that I know I will return to again and again," their wine critic Eric Asimov raved.

After eating we made plans to explore rare grapes in the region, and I was scheduled to work for a day in a community vineyard Vouillamoz co-founded, high on a mountainside.

My longest conversation with Vouillamoz took place one evening at his modest apartment, where he made raclette, a regional dish of melted cheese. I mistakenly expected something like fondue. It wasn't. In the past people working in the mountains would cut freshly made wheels of cheese in half and roast the edges over a fire. When the cheese softened it was scraped onto a slice of bread. Vouillamoz paused to take a call from his daughter, who was at a sports event, and told me a story about one of his other teenagers. "My son wants to be a chef, and I told him I would eat in his restaurant every night. He said that's fine, as long as I pay." The children often tease him, which he said is OK, since it's their way of showing love.

Vouillamoz brought out an elegantly simple modern raclette cooker. It holds a large piece of cheese under a long, narrow grill. When the time is right you swivel the base and flip it sideways, allowing the bubbling tastiness to be scraped onto a plate. "This is my favorite meal. I could eat it every day," he said. We ate the raclette with boiled, fresh local potatoes and local wine. The seared cheese takes on a dense, chewy, flavorful character as it cools, almost like a steak. It was fabulous. I had at least two servings. We drank a bottle of Diolle, a super-tart local white, then a 2012 bottle of Himbertscha made by his friend Josef-Marie Chanton. It was another crisp white, with hints of apricot and basil, and I wouldn't have traded those two obscure Swiss wines for any bottle of Chardonnay in the world.

I confessed that for many years I had assumed the French invented winemaking, and that I was also unsure when or where the first grapes appeared. Vouillamoz replied that grapes originated at least fifty million years ago, but as for winemaking, a key development happened much more recently. "The wild grapevine is a botanical species called *Vitis vinifera*. We make it a subspecies, *sylvestris*, which means it comes from the forest—'*silva*' is the

forest. It does grow naturally in Eurasia, from Portugal to Tajiki-
stan. Mainly it's around the Mediterranean Sea, or along rivers like
the Danube or Rhine. It's only approximately eight thousand years
ago that human beings first attempted to domesticate it. This is
the story I worked on a lot together with my friend and colleague
Patrick McGovern."

"The wild grapevine has significant differences compared to
what we cultivate today. Wild plants are dioecious—Greek for
'two households,'" Vouillamoz continued. "It's a botanical term
to say that some vines are 100 percent male in their flowers and
some vines are 100 percent female." But a small percentage of those
grapes, like humans, are sexually different. "If you go to a natural
wild population of grapevines, you find one or two percent of the
plants are hermaphroditic, they have both male and female parts
in their flowers."

In 2003 Vouillamoz joined McGovern on an expedition to Armenia, Georgia, and eastern Turkey, which are part of a region on
the border of Europe and Asia. Both scientists agree on a theory
of how early humans first domesticated grapes and made wine: it
started when someone observed that a few vines *always* produced
fruit. "A Neolithic man or woman who wants to take this plant
home, if he takes a male, he will never have berries," Vouillamoz
said. "If he takes a female and there is no male in the vicinity, he
will never have fruits. If he takes a hermaphroditic plant he will
have fruits every year. And this he will want to keep it, and propagate it. So this was the starting point, the first step of domestication." Today almost all domesticated grapes are hermaphrodites.

No one knows precisely where humans first domesticated
grapevines—it probably happened in several isolated parts of the

Caucasus, or around the headwaters of the Fertile Crescent (southeastern Turkey). But Vouillamoz told me about visiting the oldest confirmed winemaking site in the world, dating to thousands of years before anything remotely similar in France. It is in an Armenian cave near the village of Areni. A small stream curves around the base of a knobby mountain, where a path winds up to a triangle-shaped opening at the base of a sheer rock cliff. A narrow, sloped-wall passage leads to a larger sort of room. There, after scraping ages' worth of dirt and dung off the floor, archaeologists uncovered a three-foot by three-foot grape-pressing trough; piles of ancient grape skins, seeds, and stems; fermentation vats that held about a dozen gallons; a drinking cup made of animal horn; clay storage jars; and one of the earliest leather shoes ever discovered, a surprisingly dainty lace-up moccasin almost like a dancer's slipper. A forgotten tribe had pressed grapes in the cave 6,100 years ago.

"It was fascinating to be there," Vouillamoz reflected, since it is likely the precious drink wasn't made to be sold. "It was more a small place where you could make wine as a ritual offering to the gods. In one of the jars they found the skull of a child, which means it was also probably related to some human sacrifices." Archaeologists from Armenia and UCLA also found human bones with signs of being cut, bitten, and boiled, suggesting cannibalism, and DNA tests showed that three of the ritual victims were young women—perhaps sisters. As for the grape remains, scientists determined they were an intermediate mix between domesticated and wild grapevines.

It is also true that winemaking may have developed in that region since so many other crops and livestock were first domesticated around the Fertile Crescent, including wheat, barley, and olives, along with cattle, goats, and pigs. So even with all the evidence supporting the theory of grape domestication in the Caucasus, McGovern and Vouillamoz kept looking, and found that linguists who examined the origins of the word "wine" discovered an interesting parallel. In a broad range of languages, from ancient to modern times, "wine" appears to stem from a common ancient root word with origins in the Middle East. From Hittite to Akkadian, early Semitic to Hebrew, and ancient Egyptian to Greek, there is evidence the word *woi-no* or perhaps *wei-no*, inspired the word for wine in many other languages.

In 2013 French researchers compared DNA samples from more than five thousand wine grapes around the world. The results suggested that "domestication of [the] grapevine took place in the region spanning from the Fertile Crescent to South Caucasus, and from there spread in three directions: a Northern route, through Greece and the Roman empire to its western borders; a Southern route, through Egypt, the Arab territories all the way to Spain during the last Arab invasions; and a third route towards Asia."

The theories fit with evidence from plant and grape remains in scores of archaeological sites across the Middle East. Ehud Weiss, an expert on plant genetics at Bar-Ilan University in Tel Aviv, reviewed numerous studies and found that the earliest human settlements contain only a few grape pips amidst large amounts of lentils, wheat, and other grains. By the Copper Age, around six or seven thousand years ago, the number of grapes rose to about 5 percent of the food remains. In the early Bronze Age, a few thousand years later, grapes comprised 10 percent, and soon afterwards excavations start to show "a very large amount" of grapes. In other words, over time, people figured out how to create vineyards, instead of just foraging in the wilderness.

The DNA of domesticated plants (and animals) changed because of the choices early humans made. Vines with bigger, faster-growing grapes or sweeter and tastier fruit were favored, bitter or low-yielding ones discarded, and eventually there was a leap in productivity. That seems to have happened about five or six thousand years ago, which would help explain the rapturous wine poems and tales of the Sumerians and other early cultures. Suddenly wine wasn't just tasty. It became cheap and easy to make, a drink for the masses.

Another aspect of the wine grape's development was a complete surprise to me. Vouillamoz said that if you plant the seeds from any grape (Pinot, Merlot, Jandali, etc.) the new vine will have different flavors and characteristics, just as a child differs from its parents. Old grape varieties are essentially genetic proof that people replanted or expanded vineyards not from seeds but from cuttings, which if handled properly will take root in soil, just like the stem of a rose will. Cuttings have exactly the same characteristics as the original plant. "So if you think of an old variety like Pinot, we estimate that it's about two thousand years old," Vouillamoz said. For that whole time Pinot was propagated by cuttings.

Our famous grapes kept their general flavor profiles only because they weren't allowed to crossbreed with other varieties for a long, long time. Early winemakers latched on to the hermaphroditic wine grapes as a way of promoting consistent flowering and fruiting, then added another level of control by using cuttings to expand vineyards and lock in flavors. But some new varieties happened by chance or through viticultural roaming. For example, about five thousand years ago domesticated grapes and winemaking rapidly spread out of the Caucasus Mountain and/or Fertile Crescent regions. New tribes and civilizations explored different ways of planting and harvesting, and wild grapevines along the way gave the cloistered cuttings chances to mate, creating a viniferous melting pot. Soon the new industry spread up through Europe, often following major rivers. Vouillamoz and I were on one of the old Swiss wine routes.

Vouillamoz had studied at the University of California, Davis, one of the world's leading wine research schools, but I didn't know how he ended up there. "I was trained as a biologist at the University of Lausanne. After my graduation I made a PhD thesis on molecular systematics, which is basically the classification of plants by their DNA," he told me. "And at the same time I was a wine lover, and I was learning more and more about wine, about grapes. So I thought, why not try to combine both?" In 2001 the Swiss government gave him a grant to study at UC Davis, and the timing was perfect. In the mid-1990s Carole Meredith, a plant geneticist at UC Davis, and John Bowers, one of her PhD students, realized that the revolutionary tools for studying human DNA could be used on grapes.

To understand how radical that was, consider what came before. Early wine writing from ancient Greeks, Romans, Jews, and other cultures described grapes mostly by where they grew, if they mentioned the particular grapes at all. That means we essentially have no way of knowing exactly what variety this passage from Pliny the Elder described: "Our ancestors set the highest value upon the wines of Surrentum; but at a later period the preference was given to the Alban, or the Falernian wines." That's like saying cheese from Wisconsin used to be the best, but now people like Vermont's. It doesn't tell you anything about the kind of cheese. "They did not have the concept of grape variety that we have to-day," Vouillamoz told me. Adding to the confusion, the same grape was often called different things in different countries, in many languages.

There were early efforts to describe grapes by the berry size and color, leaf structure, or seed. The last one is quite accurate—even fossilized seeds that are millions of years old retain some distinguishing characteristics. But seeds aren't very exciting to look at. By the eighteenth century the quaint and colorful science of ampelography—a term created from the Greek words for "vine" and "writing"—was the rage. It's really just a fancy way of saying experts looked at leaves and grapes, did meticulous drawings and measurements, and decided whether one plant looked like another.

Careful ampelography using modern instruments can document important botanical differences between grapes, but like the nineteenth-century doctors who tried to gauge intelligence by measuring people's heads, it has flaws. Shape and color don't always reveal a grape's true heritage, but it was the best option for many years. Ampelography eventually inspired a small industry. Agricultural schools preserved vast collections of grape pips and leaves, then artists and publishers created lavish books and portfolios filled with color plates, the oenophile equivalent of Audubon's birds. The attraction still exists: in 2012 an Italian publisher issued a 1,500-page, three-volume coffee-table set on the history of ampelography, with 551 reproductions of classic illustrations.

Meredith's UC Davis lab helped change everything. Like lawyers

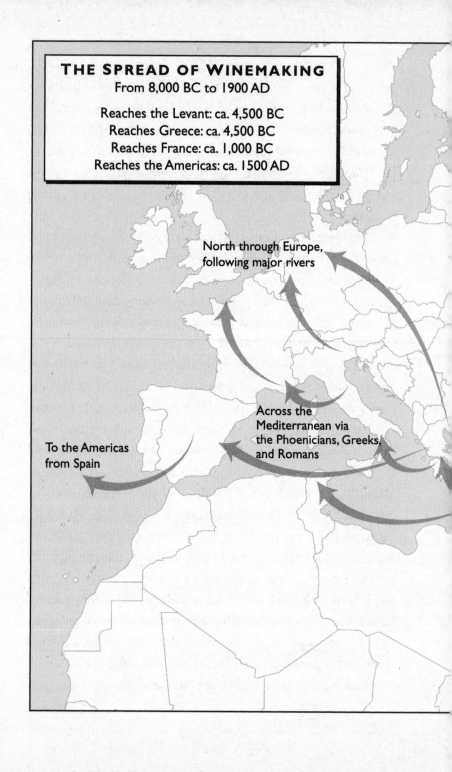

THE SPREAD OF WINEMAKING
From 8,000 BC to 1900 AD

Reaches the Levant: ca. 4,500 BC
Reaches Greece: ca. 4,500 BC
Reaches France: ca. 1,000 BC
Reaches the Americas: ca. 1500 AD

North through Europe, following major rivers

Across the Mediterranean via the Phoenicians, Greeks, and Romans

To the Americas from Spain

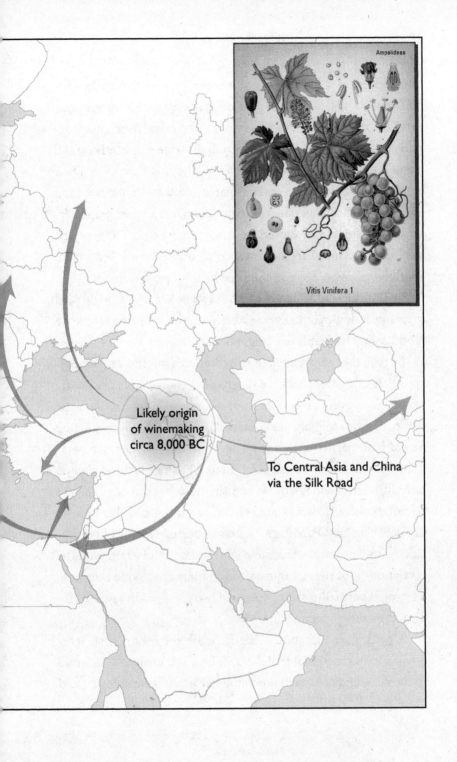

Ampelideae

Vitis Vinifera 1

Likely origin
of winemaking
circa 8,000 BC

To Central Asia and China
via the Silk Road

in paternity lawsuits, the researchers used science to dispel myths and rumors. Just as human genealogy reveals ancestors with colorful or "disreputable" pasts, or those who came from countries, religions, or races you never expected, the wine grape tests caused some major grapevine soul-searching. The first conclusive DNA evidence of a famous wine grape's parentage came in 1997, showing that the red mainstay Cabernet Sauvignon had a white grape parent, Sauvignon Blanc, and a red one, Cabernet Franc. The vineyard love affair occurred in the seventeenth or eighteenth century in southwest France. Meredith put the birth in perspective, writing, "A single pollen grain landed on a single flower and a single seed grew into a single plant. Every Cabernet Sauvignon vine across the world comes from this one original vine."

In 1999 the lab deciphered Chardonnay's origins, and discovered that one parent was the disgraced Gouais Blanc, previously uprooted or even banned in most of France. To the stuffy wine world, that was roughly like the Queen of England suddenly listing an old Irish washerwoman as an ancestor. In 1539 a German botanist noted that some people call Gouais "shit grapes," to be planted only in inferior vineyards. When the news about Chardonnay's parents broke, a *New York Times* headline proclaimed "For a Noble Grape, Disdained Parentage." Jancis Robinson, who is often considered the most knowledgeable wine critic in the world, said the revelation "unsettles all our preconceptions about vine breeding. We would normally assume you need two good-quality parents for a good-quality progeny."

A battle over Zinfandel's origins, dubbed the ZinQuest, actually led to international trade wars. The grape came to California in the 1820s and became synonymous with that wine industry, and later had a reputation as a poor man's drink. In the 1970s scientists suspected a link to Italy's Primitivo, but couldn't prove it. Some Italians got the idea of marketing their wine as Zinfandel in the United States. Same stuff, right? The Bureau of Alcohol, Tobacco, Firearms and Explosives disagreed, fueling lawsuits between the United States and the European Union that simmered for years. Carole Meredith's lab settled the issue—sort of—in 2002.

DNA showed that California's Zin was the same as Italian Primitivo, and the same as a Croatian grape called Crljenak Kaštelanski. But the earliest name for the variety was Tribidrag, dating back to the fifteenth-century Croatia, so *Wine Grapes* used that, even though all the different countries still use their own names. Meredith called the grape ZPCT, from the initials of the four common names. In the end, though, Californians didn't much care for Italy's or Croatia's Zinfandels, even though the winemakers were using the exact same grape.

Talking with Vouillamoz I got a picture of a handful of scientists doing groundbreaking research, with little if any support from the mega-wine corporations. The global market is worth over $300 billion, with total production of about 6.8 billion gallons of wine in 2016. "Everything I do in wine writing or research, I do it independently," Vouillamoz told me. He also does research on plant genetics for a Swiss government agency, but they don't let him focus on wine grapes. McGovern and Vouillamoz eventually ran out of funding for their research in Turkey. They had found a noteworthy hermaphroditic DNA sequence in a wild vine near the headwaters of the Tigris River, and hoped it might provide an important link to grapes across the region. But they couldn't continue their quest.

Vouillamoz still dreams of finishing that research, difficult as it may be. "The Holy Grail of grape domestication would be to follow the path from modern cultivars to the original vine that was initially domesticated. But we have gaps in history, we have gaps in genetics. I hope that one day, during my lifetime, we will have all the grape varieties in the world DNA profiled, enough to understand all the links. Not only for the famous European or international grape varieties, but for all of them," he told me.

I began to imagine what he was talking about—an eight-thousand-year-old family tree, with branches (or vines) twisting in all directions, each shoot representing a different variety, a new flavor. The strands of DNA in every grape hold pieces of that story. With that in mind, I told Vouillamoz how I became interested in genomics.

We hear about DNA all the time, but I hadn't really understood the details until I received a life-changing fellowship at the Marine Biological Laboratory in Woods Hole, Massachusetts. Over the course of two weeks a small group of journalists extracted DNA from surf clams. We replicated a previously published study, with constant help from grad students as well as the original scientists. After a series of steps in the lab, a small, roundish clump of pure DNA was perched at the end of my pipette. It looked like a cottony piece of taffy. At first I was just thrilled to have not botched the work, but I soon realized how DNA explains both life and history, together. It was a revelation.

DNA contains actual pieces from the past—tiny snippets of code from ancient algae, plants, fish, and mammals. I analyzed the surf clam DNA using a computer program that compares any sample with existing databases of biological sequences. The results turned up little sections of DNA that the clam shared with other creatures from both sea and land, including humans, down to the exact same combination of the letters A, C, T, G, which represent the age-old relics of the earliest life forms—nucleotides—which we all descended from.

For a while I couldn't eat a bowl of clam chowder in the same way, always thinking about our shared genomics. My experience in the lab was how I knew science could help explain the origins of the Cremisan wines that I was tracking. Although I never used that computer program (called BLAST, for "Basic Local Alignment Search Tool") to compare human and grape DNA, others have, and some sections of DNA match. (Which makes sense—after all, scientists at the National Human Genome Research Institute say about 60 percent of our DNA is also in bananas.)

After we finished dinner I looked around Vouillamoz's apartment. It was clean and organized, with a refrigerator-sized temperature-controlled wine cellar near the front door. I asked him how grapes manage to produce so many flavors. He admitted that particular field wasn't his specialty, but gave a few examples anyway. "Every grape has its own aromatic profile, and even within the same grape variety you observe some differences. A very famous example is Gewürztraminer," he said. I knew little about the pink grape widely grown in Germany and France, though I had enjoyed the wine itself. Vouillamoz said Gewürztraminer is the same as Traminer; the French call it Savagnin Blanc, which is a very old grape variety. But one vine had an aromatic mutation in its DNA and developed an aroma of lychees. All the Gewürztraminer in the world can be traced back to that one mutation. If a rare grape variety dies off, that flavor profile dies with it. Of course yeast, soil, and winemaking expertise all influence the character of a wine, but grapes are the raw material.

Other unseen forces can change flavors, too. Some believe that vines with lower controlled yields produce more flavorful juice, and in any case productivity is a key aspect of any vineyard. Some Gamay vineyards have granite-rich soils where a species of nematode thrives. The tiny, worm-like, almost parasitic creatures become vectors for a virus. "These Gamay vines are attacked by this virus—but not too much. Just enough to calm it down and to make it produce much less. So it's a natural yield regulation that makes the Gamay wines better," Vouillamoz said.

Part of the reason Cremisan's wine captured my attention was because all the shelves and restaurant lists seemed packed with endless bottles of Chardonnay, Merlot, and the like. I knew about the industrial farming angle of those popular varieties, but Vouillamoz described an even more complicated story. For most of winemaking history, the typical vineyard differed greatly from what we see today.

"Until the end of the nineteenth century, traditional vineyards were planted in what the French call *en foule* ['in a crowd']. Which means that in the same vineyard you have five, eight different varieties. Some would harvest at different times, some would harvest everything together," he said. Skilled winemakers used the diversity to enhance flavors. The early grape varieties balanced out the more complex late ripeners, and sweet varieties balanced acidic ones.

I knew that phylloxera (a tiny, aphid-like pest) had destroyed vineyards throughout Europe starting in the 1860s, but never thought about the impact that had on diversity. Two things happened during replanting. "First, they had to choose which variety to use, and in many cases—not only in France, but also in Switzerland, Italy, and Germany, and everywhere—they had a tendency to forget the old grandfather varieties. And they chose to plant varieties that were easier to cultivate, and especially that would produce more. So that's why in many regions some of the ancient, traditional varieties have been almost abandoned, or sometimes have disappeared," Vouillamoz said.

"The second thing is they started to plant vineyards at a large scale. So they needed to be rational. They needed to make one line of Cabernet, or one vineyard of this or that. Vineyards became regimented, all vines aligned, everything is perfect, nothing in the middle. And if you just think about that, since this new way of planting a vineyard, if you have a chance seedling falling in the middle, such as Cabernet Sauvignon, it will not survive." The new way of business radically lowered the chance for new grapes to take root.

In the 1870s vineyards started using pesticides and other chemicals in large quantities. The goal was to kill the phylloxera and various diseases, but many other insects and soil microbes suffered, too. That wasn't all. The notion—and the marketing—of so-called "noble" grapes was so successful that vineyards all over Europe uprooted old, distinctive varieties. Much later, subsidies from the European Union funded vast campaigns to uproot vines via a program that in retrospect had a sinister name: grubbing up schemes. From the 1970s until recently, millions of acres of vineyards were "grubbed up," mostly in France and Italy.

The idea had some merit—Europe was producing so much wine that millions of gallons failed to sell each year and were turned into industrial alcohol via another expensive program. But there were nasty side effects. Many older vineyard owners took permanent buyouts, and small plots of rare local grapes disappeared. During the same era international wine consultants and multinational corporations preached a related gospel of modernization: grow the supposedly more profitable Merlot, Chardonnay, and Riesling. That's what customers want, they said.

Vouillamoz reflected on how the world of wine has changed over the last twenty years. Decades ago his friend Giulio Moriondo fell in love with rare Northern Italian wine grapes, and was scoffed at. "When he was working at the viticulture institute he got crazy about local varieties and people said, 'Come on! This is what our grand-ancestors would grow! It's not interesting. Let's try new things like Chardonnay, Cabernet.'" Moriondo said no, the old varieties are fascinating, and he catalogued dozens of local grapes. His superiors still weren't interested, so in 2001 he quit the job.

The institute destroyed Moriondo's whole collection.

"Fifteen or twenty years of gathering of material," Vouillamoz lamented. "So, at one point in history you want to do something for your region, and people don't understand."

"Tragedy," I said. "Ah, people."

"Now, twenty-five years later, they want to start again, out of nothing. That's a shame."

Vouillamoz said many winegrowers who express interest in rare grape research never follow through, focusing instead on ways to increase production. Others complain about DNA results. In 2006 he discovered that Sangiovese, long claimed by Tuscany in northern Italy as the premier grape used for Chiantis and Brunellos, had a parent from Calabria, a southern region near Sicily.

"So when you started uncovering this different history, were people welcoming, or resistant?" I asked.

Vouillamoz laughed. "I can tell you that Tuscan people were upset." They were unhappy about the Calabria link because they look down on southern Italy. In fact, wherever DNA challenged established narratives, wine producers had strong reactions. "By strong I mean emotional. Because they really would like a given variety to be their own variety, and that's it. It's like you take their babies out of their hands."

Over time, though, most winemakers accepted the news. "The same people who had, let's say, bellyached, discovering that their variety did not originate from their region but from the region next door or from southern Italy instead of central Italy, a few years later they would tell me, 'Well, thank you. Because now we have another story to tell about our wine, and we sell more. We sell much more.' It's the storytelling—wine marketing. It's nothing else," Vouillamoz said.

Despite all the challenges, Vouillamoz sees much recent progress. In the late seventies and early eighties some Italian winemakers weren't proud of their local grapes, and were almost ashamed to use them. "But they would be proud to say, 'Oh we have Cabernet, we have Chardonnay.' Then they realized they are the only ones to have the indigenous varieties." Now, after the publication of *Wine Grapes*, people know there are many varieties of grapes to play with, and to plant. He half-joked that I could help inspire a revival of the Mission variety in America, referring to grapes the Spanish brought to the New World centuries ago.

I told Vouillamoz that he and McGovern seemed to have a pretty firm grasp on the origins of wine. Instead of accepting the praise, he corrected me and cautioned that the research was still in the early stages. For example, while there is now a fairly solid understanding of how the leading European grape varieties relate to each other, the likely "cradle of grape domestication" is a different story. "For Georgia, for Armenia as well, no one knows exactly which are the 'founder grape' varieties of these areas," he said. The same is true of Israel and the Holy Land, which helps explain the uncertainty about my Cremisan grapes, and where Drori's research fits in.

The *Wine Grapes* authors came up with the term "founder grapes" to describe the ancestry of most famous European grapes. "We realized that a limited number of varieties are responsible for most of the diversity that we observe today," Vouillamoz told me. One diagram in *Wine Grapes* shows 156 grapes that are related to each other, and the little-known Gouais Blanc is now considered a key variety for European viticulture. "I call it the Casanova of grape varieties, because it had children in every country it visited," he said with a smile.

My Swiss vineyard visits were just as illuminating as the talks with Vouillamoz, in very different ways. On one clear, crisp morning I found myself high on a mountainside. The Swiss Alps looked grand. I didn't. A fine, smudgy, cough-inspiring dust covered my hair, hands, and clothes. My back hurt, my feet ached, and I was exhausted from lugging large rocks and buckets of smaller ones to help build a vineyard terrace, at an altitude of about three thousand feet. Wine magazines always show pictures of the lovely vines, the heavy bunches of grapes, the stately oak casks. Sometimes a neat or rustic stone wall appears in the background, but they never show the stone-schleppers.

As the knowledge of winemaking spread west from the Caucasus, it also moved north. Major rivers turned out to be ideal wine grape highways. Along the Volga, Don, Danube, Rhone, Rhine, and Loire, winemakers found receptive climates and soils. The imported grapes mixed with the wild local stock and flourished. I'd learned all that from experts, but they failed to mention how hard it is to go from the bottom of the mountain valleys, where the rivers are, to the vineyards far higher up. At least it was hard for me.

I'd come to Switzerland armed with audio recorders, notebooks, and a camera, but unprepared for vineyard work. So perhaps I was due. I was at VinEsch, near the town of Visp, ten miles from the Italian border. It's a tiny, nonprofit, rare grape sanctuary, supported by Vouillamoz and other wine lovers. In the 1970s a winemaker named Josef-Marie Chanton found a few vines at the VinEsch site, and he propagated them for replanting. Himbertscha was one, a white grape headed towards extinction. At the time no one else seemed to care. In 2010 the vineyard's elderly owner, in poor health, decided he couldn't continue. The vineyard would be sold or abandoned. Chanton called Vouillamoz and told him, "We cannot let it go. We should do something." They met, decided to try to buy the vineyard, and quickly found thirty-three sponsors willing to put up about five hundred dollars each. Eventually the membership doubled.

Vouillamoz and Chanton were amazed to see so many people from all walks of life respond to the call. In other words, rare grapes mattered, and not just to the scientists. "People from Bern, or Zurich—it is two or three hours by car—come during the weekend just to help for free. Because they are proud to be part of a project to rescue an old vineyard," Vouillamoz told me before I made the climb. Now Chanton smiles and tells people he makes the best Himbertscha in the world. Well, yes, because he's the only one who makes it.

The morning wore on at VinEsch and I took a break to sit on the long, ten-foot tall terrace I'd played an exceptionally minor part in building. A tiny rivulet bounced downhill along the edge of the vineyard. I looked across the valley and admired the hundreds of terraces built by generations of stone-haulers. All for wine. The scale of the effort astounded me, now that I knew a little about what it took. Despite my aches I was proud to be helping the volunteers at VinEsch. There are more than a dozen terraces in the vineyard, and while many of them are over a hundred years old and only need maintenance and repairs, the association built several new ones from scratch under the guidance of a master stonemason. My group filled in behind the walls of the new terraces to create flat ground for new plots of grapes.

We took a break for lunch, drinking local wine and eating sandwiches at a picnic table on a spot of flat ground, next to a small old shed built of rough-cut lumber, its roof covered in large, natural pieces of slate. The group included Philippe, a Swiss journalist; Max, a retired scientist; Bettina, from Bern; and Christoph Holderegger, who told us there had been snow there until recently. Now tiny light purple flowers bloomed along the stone and gravel paths at the edges of the vineyard. Holderegger said larger groups of volunteers turn out later in the season.

After lunch I dragged more stones. My energy faded. It was early spring, chilly, and I didn't have a sweater. I shuffled over to the shed and went inside. The room had a low ceiling, a little window, and a narrow bench just long enough for me to lie down on.

I had expected to spend a lot of time interviewing the VinEsch members, but they were focused on working, which explains why the trains actually do run on time in Switzerland. I looked out the window and saw snow-covered peaks in the distance. The small road at the bottom of the valley was so far away that the occasional cars that drove by looked like toys.

My extra break was embarrassing compared to the diligent Swiss, but I couldn't help it. Sure, some wine writers squash a few grapes with their feet to see how it feels, or even get up at dawn to pick dew-covered grapes during harvest. A few even try to make wine. Rock-hauling is not a normal story line, nor are weeding, bug-killing, pruning, filling bottles, packing cartons, or all the other tasks that go into bringing us bottles. Eventually I got a second wind, left the shed, and went back to the rocks.

After all the talk about the origins of wine I'd heard, the stone-hauling made something else click. In prehistoric times birds and other creatures spread the seeds; later humans took over, creating new habitats like this remote Swiss mountainside. We did more than drink wine for thousands of years—we helped grapes evolve into new and sometimes tasty varieties.

On another day Vouillamoz showed me how a village postman tried to preserve a grape variety no one else wanted. We visited Bovernier, population 875. The valley was bursting with fresh green colors and a nearby stream surged with a taupey, murky color. The Swiss have a name for that: glacial milk. Slowly but inevitably, the huge fields of ice in the mountains crush the rocks underneath to a fine power, and it washes down to streams, giving them a shimmering opacity.

We had come to Bovernier in part to taste a new vintage of
Goron wine, a forgotten red grape. The village postman, the late
Roger Michaud, had championed it. "He was the only one to really believe in this variety when all the other people in the village
were saying, 'Are you crazy? There are much easier varieties to
grow, why do you bother with this one?'" Vouillamoz recounted.
After Michaud died in the early 2000s, Goron grapes almost vanished, too.

We met Chantal Sarrasin, a proud local official, and she explained that around 2012 villagers were looking for ways to expand
tourism and the local economy. They decided to use public funds
to create a communal vineyard. Some suggested planting Goron,
the historical variety of the village, while others said it was a bad
idea, since older generations must have stopped growing it for a
reason. The rebirth came after one person remarked that they had
essentially killed off their own heritage, since historical documents
described Bovernier as the land of Goron. An enologist (a winemaking expert) endorsed the idea, and the village bought plots
covering about fifty acres of a nearby hillside.

Sarrasin and a few other leaders of the effort gave us a quick
tour of the neat, trellised vines, and then we headed back to the
village to try the 2013 Goron, the first vintage from the new plantings. Along the way we passed an elderly white-haired woman,
dressed in dark pants, a bright pink sweater, and a cream-colored
quilted vest. Our car stopped suddenly. After a few excited words
of introduction from Sarrasin, Vouillamoz burst into smiles. It
was Cecile "Fifi" Michaud, the widow of the postman who had
defended Goron grapes when no one else would. She was ninety-
two years old.

Vouillamoz tilted his head as he listened intently to every word Cecile said, while also eagerly asking her question after question, gesturing with his hands. She beamed at the attention, laughed gently, then openly, and held her hands out, palms up, describing her late husband's passion for the obscure grape. Vouillamoz had talked to him by phone years before, but they never met. "This is amazing!" he told me, sounding like a teenage boy who meets the Hollywood starlet of his dreams.

We all went back to the village to try the latest bottle of Goron, and Mme. Michaud brought out one of the last remaining bottles of her husband's 1991 vintage. In a quiet room on the second floor of a stone building, José and Mme. Michaud sat next to each other, and they couldn't stop talking. Vouillamoz admitted, "I'm supposed to go meet my children, but now this takes precedent."

The 2013 Goron had a nice, full taste of ripe cherries, and I thought it was a fine, promising wine. Mme. Michaud said her husband often let bottles age, so we were all looking forward to the 1991 vintage. She posed for pictures with Vouillamoz as they proudly held the old bottle like a baby. It had a label with hand lettering and a priest or king-like figure in the corner. Finally it was time to taste. Vouillamoz poured and sniffed the glass. His smile faded.

"Corked," he said, using the term for a bottle with a damp, moldy smell and taste, often from incredibly tiny amounts of the fungal compound TCA (2,4,6-trichloroanisole). One to five percent of wines suffer from the defect, creating losses of ten billion dollars per year for the industry. It is so distinctively unpleasant that just a few nanograms (a billionth of a gram) per liter can ruin a bottle. Vouillamoz wouldn't even let me taste the wine.

He was philosophic. "We don't care that the wine is corked.

We opened it together, and that means something," he said of the whole experience of meeting Mme. Michaud. We soon learned she wasn't about to let one setback wreck the day.

Mme. Michaud invited everyone to see her husband's former vineyard. I expected another picturesque Alpine view. After a short drive through the valley the car stopped in a nondescript parking lot. We climbed up a slope and across a railroad track. A tiny vineyard hugged the side of the mountain, protected by a rusty gate and a padlock.

The postman had obsessed over his rare grapes here. The soil, if you could call it that, was shards of rocks, smaller pieces of the shards, and rock that had dissolved into a fine gray dust. Roger Michaud's lifelong love, after his wife, was the type of place that is often desolate or trash-filled in other countries. He took the tiny plot of land, built small terraces, and created a vineyard in the bottom of a valley without direct sun much of the day.

The vineyard was barely ten feet from the tracks, and extended perhaps thirty feet in towards the mountain. The small terraces had neatly stacked and fitted rocks, and a wobbly path of stepping stones wound up the mountainside until it hit a sheer rock wall, at least one hundred feet high. At the top of the vineyard a small plateau overlooked the valley. Everything was covered in late afternoon shadows, framed by blue sky and green mountainside. I took a picture of Vouillamoz standing next to the old man's rough shed. He had a euphoric smile, as if we were touring a *premier cru* vineyard in France. A faded local newspaper article about the vineyard was still tacked to the shed's rough wood, with a picture of it back when it was perfectly tended. The headline read "*A fin que la terre ne meure,*" which means "So that the land does not die." Michaud had taken a place no one else wanted, turned it into a vineyard, and inspired the villagers with his story.

How many other small vineyards had disappeared over the last fifty years. Thousands? Tens of thousands? Vouillamoz said that even in prosperous Switzerland young people who inherit vineyards increasingly lease them out to large corporations. In our modern world it's hard to make a living as an independent vintner, and few people want to spend a large portion of their free time tending grapes. I get that now, after hauling stones in the Alps.

TASTINGS

Switzerland has many fine winemakers, but for years they didn't bother much with exports. That's beginning to change. Here are a few Swiss wines made from distinctive local grapes. If you can't find these, look for something from Valais, the region where Vouillamoz lives.

Kellerei Chanton, Valais

Himbertscha (white)

Lafnetscha (white)

Simon Maye & Fils, Valais

Syrah

Romain Papilloud, Cave du Vieux-Moulin, Valais

Païen de Vétroz (white, from Savagnin grapes)

And here is a rare wine Vouillamoz uncovered in another country:

Branimir Cebalo, Croatia

Grk (white, from a rare grape with all-female flowers)

6

Flavor, Taste, and Money

I rather like bad wine, said Mr. Mountchesney,
one gets so bored with good wine.
—Benjamin Disraeli, *Sybil*, 1845

As I tracked down clues about Cremisan's wine, a more basic question emerged: What had I really tasted? I don't mean the type of wine, or what grape variety it was, though I thought of it that way at first. I mean how my brain processed the wine. The latest research suggests that even before that first sip in the hotel room, my preconceptions influenced what we call taste.

Neuroscientist Gordon Shepherd thinks our historical understanding of flavor gets some key points wrong. He writes that flavor is not in food; it is created from food by the brain. Shepherd, a professor at the Yale School of Medicine, received his MD from Harvard and his PhD in philosophy from Oxford University, so he brings an entirely different rigor to the question of whether a bottle is "plonk," to use the slang term for barely drinkable wine, or some critics' vision of perfection—a 100-point score.

Shepherd and others use real-time brain scans to study how the mind reacts when people sip wine or eat potato chips, and different areas do light up with activity. In a 2015 journal article entitled "Neuroenology: how the brain creates the taste of wine," Shepherd found that multiple brain regions fire up before, during, and after we drink a glass, "apparently engaging more of the brain than any other human behavior."

He's hedging on the claim, but the data is impressive, with several key points: Shepherd said our experience of flavor begins before we open our mouth. "The first step . . . occurs entirely in the head, consisting of the accumulated experience of the taster with wine in general and anticipation of this wine or wine tasting in particular. *The expected flavor of the wine is thus due entirely to vision and to the imagination.*" (Emphasis mine.)

Shepherd's theory suggested how preconceptions might have influenced my first Cremisan experience. In that hotel room I was tired, the label surprised me, and some part of me was curious even though I usually avoid mini-bars. What synapses fired up in my brain that evening? I drank wine for many years before thinking about any of these points.

Of course, Shepherd doesn't claim that flavor is *all* about preconception, just that many senses are at work, including sight (how wine looks), sound (fizzy bubbles or even background music), and aroma (which accounts for most of what we call taste). We don't just smell wine by swirling it in a glass and sniffing. After swallowing, a secondary burst of aromas travel from the throat into our nasal passages, and without that we'd miss a tremendous amount of flavor.

In 2014 French researchers monitored ten expert sommeliers and ten non-experts of roughly the same sex and age as they each tasted a red wine and a white wine. The MRIs found some common areas of brain activation, and differences. Everyone's insula, operculum, and orbitofrontal cortex lit up—areas that help process tastes and odors. But the wine experts also showed brain stem activity, suggesting that they immediately start cross-referencing various types of memories. In other words, all the training may really change how their brains experience wine. "A sommelier can distinguish a subtle difference of taste in wine by training their ability to integrate information from gustatory and olfactory senses with past experience," the research found.

That doesn't exactly mean somms are better tasters, since other studies question how well people really remember flavors and aromas. But somms do put their experience to use, and the new research illuminates why people respond differently to the same food or wine. Taste buds vary widely. So when you go bonkers over a particular vintage and share the news, there's a chance it could be like praising Rembrandt to someone who prefers de Kooning.

If you're enjoying a glass in a wine bar or buying a bottle in a store, take a moment to consider the background music. In one field study a display of French and German wines was accompanied by music that alternated between the two countries. The results? Playing French music increased sales of that country's wines, but German music had the opposite effect—sales of French wine went down. The customers weren't aware of the musical influence in their purchasing decisions, and the researchers remarked that the study raised ethical questions about in-store music.

In another study twenty-six people tasted three different wines, with and without classical music. They "perceived the wine as tasting sweeter and enjoyed the experience more while listening to the . . . music than while tasting the wine in silence." The same researchers studied twenty-four people while pairing specific wine with specific music. The overall consensus was that Mozart's Flute Quartet in D major went better with white wine, while Tchaikovsky's String Quartet No. 1 in D major paired better with red.

But what about the grapes themselves? Flavor and aroma aren't all mind games. A 2007 paper by French and Italian scientists in the journal *Nature* found that grapes have many more flavor-producing genes than their close relatives, poplars and rice. For example, the grapevine has eighty-nine functional genes that drive the production of terpenoids—types of resins, essential oils, and aromas—compared with poplar's thirty or forty. The research also found that wine grapes have forty-three genes that drive the production of resveratrol, which may have health benefits. The scientists concluded that it may someday be possible to trace diverse wine flavors down to the genome level.

Laboratories can now detect incredibly tiny amounts of flavor. In 2008 Australian researchers identified the precise compound that gives Syrah wines their peppery taste. It's called *rotundone*, and in grapes it's present only in the skin (it's also what makes pepper peppery). Some people can detect rotundone concentrations as low as two parts per billion in red wine (that's roughly the equivalent of one drop in a nine-thousand gallon liquid tanker truck).

Consider this: about 25 percent of people can't taste rotundone at all, even in much higher concentrations. Jamie Goode, the British wine writer, tells of one professor who breaks up lectures on flavor by handing out strips of blotter paper treated in a particular solution. When students put the paper in their mouths, about a quarter taste nothing at all, half find the taste bitter, and another quarter experience an extremely unpleasant, intense bitterness. In other words, there are biological (and probably genetic) differences in how we taste. Scientists call the most sensitive group "super-tasters."

Continuing with my earlier analogy about painters and personal preferences, trying to share your love for a particular wine could be like showing a Matisse to someone who is color-blind. About 35 percent of American women are super-tasters, compared to just 15 percent of men. (Yet more men are color-blind, so I'm not sure where that leaves us in making comparisons!)

Take a moment to consider the implications of all this variation for wine lovers—and wine critics. No matter how carefully or honestly someone describes a taste, a significant chunk of the population will experience the same bottle in a different way. Not only that, but money changes everything, as the song goes.

A Stanford University study offered the same wine to people twice during blind tastings—telling them one bottle cost ninety dollars, and that the other sold for ten. The supposedly higher-priced wine was deemed to be better. There's also this story: A sommelier found an enticing French rosé for just three dollars a bottle. He priced it at three dollars a glass, ten per bottle, feeling he was doing customers a favor. The cases gathered dust. The owner scowled. The somm couldn't give the wine away, until he raised the price to seven dollars per glass and twenty-eight dollars per bottle.

It immediately became one of the most popular wines on his list, and six cases sold out in a week.

In another experiment people were served the same white wine twice. The only difference? One bottle had a California winery label, while the other said the wine came from North Dakota. The test subjects overwhelmingly rated the "California" wine as far superior. In one more famously sneaky test, researchers again served identical white wine twice, but the second round had flavorless red food color added. The drinkers described the colored white wine as if it were a red!

Then again, instincts may sometimes be right. Steven Shapin, a History of Science professor at Harvard, points to a passage from *Don Quixote* that manages to both poke fun at and honor wine-tasting expertise. In the story, villagers doubt the quality of a hogshead, or barrel, of wine and ask Quixote's squire, Sancho Panza, for an opinion. Sancho tells them:

> ... I have had in my family, on my father's side, the two best wine-tasters that have been known in La Mancha for many a long year, and to prove it I'll tell you now a thing that happened to them. They gave the two of them some wine out of a cask, to try, asking their opinion as to the condition, quality, goodness or badness of the wine. One of them tried it with the tip of his tongue, the other did no more than bring it to his nose. The first said the wine had a flavor of iron, the second said it had a stronger flavor of cordovan [leather]. The owner said the cask was clean, and that nothing had been added to the wine from which it could have got a flavor

of either iron or leather. Nevertheless, these two great wine-tasters held to what they had said. Time went by, the wine was sold, and when they came to clean out the cask, they found in it a small key hanging to a thong of cordovan . . .

I did find something to celebrate amidst all the evidence of marketing and psychological pitfalls. Lesser-known wines had an unexpected bonus: there's little to no "hype markup" on forgotten grape varieties (at least not yet). Even with a focus on small vineyards that produce handcrafted wines, I wasn't paying a premium—a refreshing contrast to the astounding cost of famous French, or more recently, California wines. Most of the wines I enjoyed cost between fifteen and forty dollars, which seems very fair for an uncommon experience. If you don't love them, it wouldn't be a great loss.

The prices of the famous classic vintages do present a real problem. The vast majority of wine drinkers can't afford the stuff and never have a chance to taste it. Consider Burgundy's Domaine Romanée-Conti. Often portrayed as one of the best wines in the world, a single bottle could theoretically cost you thousands of dollars. Theoretically, because you can't just buy one. New vintages are doled out via an opaque process that rewards long-standing distributors. You can put your name on a list, try to buy a bottle at auction, or even gamble on what might be called the wine futures market, where you pay huge sums of money years in advance in hopes of getting a bottle.

Perhaps the French poet Charles Baudelaire had the right attitude in the 1850s: "Wine resembles man. We will never know how far it is to be prized or scorned, loved or hated, of how many sublime actions or monstrous crimes it is capable. Let us not then be more cruel towards it than we are towards ourselves, and let us treat it as an equal."

All the modern studies about wine made me consider what judgments ancient people made about it. There's plenty of evidence that quality mattered. Pliny the Elder wrote this about the differences in wines almost two thousand years ago: "Who can entertain a doubt that some kinds of wine are more agreeable to the palate than others, or that even out of the very same vat there are occasionally produced wines that are by no means of equal goodness, the one being much superior to the other, whether it is that it is owing to the cask, or to some other fortuitous circumstance?" Another line echoes a dilemma we still face: "I shall not attempt, then, to speak of every kind of vine, but only of those that are the most remarkable . . ."

The Greek writer Athenaeus might have qualified for a job with Robert Parker's influential *Wine Advocate*, based on his exuberant prose: "As for Magnesia's sweet bounty, and Thasian, over which floats the smell of apples, I judge it far the best of all wines excepting Chian, irreproachable and healthful. But there is a wine which they call 'the mellow,' and out of the mouth of the opening jars . . . comes the smell of violets, the smell of roses, the smell of hyacinth."

There is also evidence that humans developed taste preferences at an early stage in our evolution, long before grapes were even domesticated. After leaving Israel I called the Gath archeologist Aren Maeir with some questions about our sensory systems. He told me that even hunter-gatherer societies show distinct food preferences.

They don't just wander all over the place, gobbling up everything in sight.

Maier mentioned studies of the Bushmen in southern Africa. "They have a very wide range of food that they can hunt or gather, but they choose of that wide spectrum only the very specific things they find appealing. The others they'll only use in extraordinary circumstances, for example when the preferred foods are not available, or there's a drought."

Maeir said recent excavations in Israel point towards discerning ancient palates. In 2013 archaeologists found a 3,600-year-old wine cellar at Tel Kabri in northwest Israel, three miles from the Mediterranean and close to the Lebanese border. The 66,000-square-foot complex is the largest palace from that era ever found in the country, and was likely home to powerful regional rulers. There was wine residue in forty large storage jars, each of which held fifty liters. Most of the wines were red, and they were flavored with various combinations of honey, cedar oil, juniper, and perhaps mint, myrtle, and cinnamon. "These additives suggest a sophisticated understanding of the botanical landscape and the pharmacopeic skills necessary to produce a complex beverage that balanced preservation, palatability, and psychoactivity," concluded a team of researchers from several universities. In other words, the rich wanted variety and specific tastes, just like today.

Wine "labels" turned out to be far older than I expected, too. Swedish researcher Eva-Lena Wahlberg did her master's thesis on Egyptian wine-jar labels, analyzing 444 of them, including those found at a palace city, in King Tutankhamen's tomb, and at a laborer's village.

EGYPTIAN WINE "LABELS"

Hieroglyphs contained details of specific vineyards, types of wine, and even the names of individual winemakers. Vineyards marked different jars with designations such as wine for offerings, wine for taxes, and wine for merry-making. One "label" gives these details:

Year 5

Sweet wine of the Estate of Aton

Vineyard supervisor Ramose

Others mention vineyard supervisors Hatti, Tju, and Hori, and "the vineyard of the Temple of Millions of Years . . . in the Estate of Amun that is on the river of Usermaatre-Setepenre," and the "Very good [wine] of the Estate of Aton from the Southern Oasis." Reading all those details, I could imagine some long-dead Egyptian forming a mental attachment to a particular vineyard, just as I did with Cremisan.

Reflecting on wine flavor and aroma, I came up with this list of influences: the variety of grapevine; the age of the vine; the soil it's planted in; the rainfall that season; the temperatures and periods of drought; the microbes in that soil; the nematodes or other creatures that prey on the vine; the viruses or other tiny creatures inside the nematodes; the time of harvest; the type of rock, clay, wood, stainless steel, or plastic that the grapes were pressed in, fermented in, or stored in; the temperatures when any of that happened; the way the amphoras, barrels, or bottles were sealed; how long they were stored; the emotions of the people who drank the wine; the DNA of the people who drank the wine; their individual taste buds; what they read or heard about the wine before they drank it; whether they knew how much the wine cost; how it was poured or decanted; the music and light in the room; and perhaps some other intangibles we don't know about yet. All a reminder that I still needed to look into the role of yeast and fermentation. My to-do list grew even as I checked off items.

Taking wine diversity and human diversity into account, both madly passionate wine lovers and the legions of casual drinkers make sense. Super-tasters are natural candidates for the first group—of course they go wild over all the subtle variations. For many people with average taste buds a pleasingly simple Chardonnay or Merlot may do the trick just fine. It still delivers the primordial alcohol buzz, and some can't taste all the flavors anyway.

7

The Caucasus

You are a vineyard newly blossomed.
Young, beautiful, growing in Eden,
You yourself are the sun, shining brilliantly.

—"Thou Art a Vineyard," Georgian chant, ca. 1100 AD

I stood near the banks of a large river. It tumbled out of the mountains, milky gray with flecks of granite and quartz worn off the snow-covered rock. Clouds and mist obscured the highest peaks. The ochre tower of Alaverdi Monastery stood nearby, surrounded by high stone walls silhouetted against early spring fields. A *Game of Thrones* episode could be filmed here, with marauding tribes surging across the no-man's-land. I was in the Republic of Georgia and the Caucasus Mountains, a wild, sometimes violent region home to Anatolian leopards, bears, wolves, lynx, and golden eagles. The Holy Grail of wine grapes that José Vouillamoz hopes to find could be near here, too. The most likely search area for the mother of all modern vines, if you can call it that, stretches from southern Russia down through eastern Turkey, Armenia, and northern Iran—about two hundred miles wide and five hundred miles long on maps. Reading scientific papers about the Caucasus is one thing; seeing them is another. I now understood the real-world challenges scientists are up against in deciphering the wine grape family tree.

The Caucasus Mountains rise to more than fifteen thousand feet, with glacial lakes and semi-tropical valleys hidden throughout. It is untamed land, but also a botanical and human crossroads for Central Europe, Central Asia, and the Middle East. The headwaters of the Tigris and Euphrates Rivers begin in the foothills of the Caucasus in eastern Turkey, then the rivers flow down into the Fertile Crescent, home to the first large civilizations such as Babylon and Ur. The Chechnya border was thirty-five miles north of where I stood; Armenia sixty miles south. Though Georgia is mostly politically stable of late, in January 2017 the US State Department had this to say about those nearby parts of Russia: "**Do not** travel to Chechnya or any other areas in the North Caucasus region. If you reside in these areas depart **immediately**." (Emphasis in original).

Alaverdi is in Georgia's Kakheti province, and a nomadic people called the Kush live in the nearby mountains. Anthropologist Florian Mühlfried spent time with them and found that outsiders struggle to even grasp their religion. "We are Christians. But we also worship stones," an elderly Kush woman said, as if that explained everything. The stones she referred to are mountain shrines, according to Mühlfried, who is from Germany's Max Planck Institute for Social Anthropology. The Kush, who herd sheep, hunt, and forage for wild plants, believe that ogres controlled the region in primeval times, until an army called God's Children expelled them.

The stone shrines honor the liberators. Mühlfried wrote that men still visit the shrines during summer festivals, when the immediate vicinity looks like a battlefield. "Cut-off heads of sheep and goats, sometimes tossed away over the shoulder with a small prayer . . . lie around next to carcasses in puddles of blood." Offerings of wine and pastries pile up nearby, and a common Kush toast before drinking wine is "Without our shrines, we would be lost." Women aren't supposed to approach the shrines, especially during their periods. Reading about the sacrifices reminded me of the story Vouillamoz told me about the six-thousand-year-old Armenian cave winery, which is only a few hundred miles south of Alaverdi.

It takes a little over an hour to get to Alaverdi from Tbilisi, where I was staying. Along the way my driver stopped at a little roadside farm. An old woman peered into a round brick oven built into the ground. A wooden board was propped against the top, filled with small rows of yeasty, rising dough. She slapped them on the sides of the oven, peeled them off after a few minutes, and piled them in a fragrant row. I bought one, along with some of her homemade cheese, and savored the chewy, salty tastes.

The monastery's round, 170-foot-tall, cone-topped turret is visible from miles away. The first church was built there about 1,500 years ago to coax the pagan mountain tribes into a new religion, and it kind of worked, except for the worshipping stones part. For almost a thousand years the turret was the tallest in the country; recently a newly built church went higher. Stone walls, fifteen feet high in some places, encircle the entire monastery, and the main entrance looks like a castle built to withstand sieges, with narrow slots for archers to shoot out of. It was all built out of necessity.

Roman and Muslim armies invaded Georgia, followed by Genghis Khan. The Persian shah Abbas I invaded numerous times in the early 1600s, reputedly camping in the Alaverdi church for a month and defiling it. During the Ottoman Empire the huge Christian icons on the monastery walls were painted over. In 2008 Russia invaded parts of Georgia.

I made the trip because Alaverdi still practices an early style of winemaking. Georgians use whole bunches of grapes, including the stems, to make wine and then bury it in the earth in large clay jars called *qvevri*. Qvevri may have inspired the Egyptian, Greek, and Roman amphoras that Mediterranean peoples used to make and store wine. Wooden barrels didn't come into widespread use until about two thousand years ago. Just as oak barrels impart flavor into the liquid they hold, clay qvevri give wines a distinctive, earthy taste, and also change how they age.

Up close, Alaverdi's walls are a mosaic of burnt umbers, gray washes, and faded copper highlights. I walked through an arched entrance and came into a huge courtyard. It was a mix of ruins, rebuilding, and crisp new order, with young fruit trees beginning to grow in the grassy sections. Bright green moss and lichen grew in spaces between the stone walkways. A monk named Father Gerasim greeted me. He had a dark brown beard, wore the black garb of the Orthodox Church, and spoke with a calm focus. As we walked along a path he pointed to a small plot of grapevines.

"Now it has been planted with one-hundred-and-four different varieties," he said, adding that the monks chose grapes that most wineries had forgotten. "It looks quite easy and simple, but that's not so. Because everything starts with taking care of the vineyard. We do not use any pesticides or chemical additives." But he resisted some of the labels that have become fashionable in international winemaking in the last ten to fifteen years. "We are not the followers of the natural wine, nor the biodynamic wine styles. We are the followers of the traditional Kakhetian Georgian winemaking technology—which was the first."

"Natural" generally means farming without pesticides and making wine with no additives, except perhaps a touch of sulfur to guard against spoilage. "Biodynamic" is similar to organic farming, but with an added philosophy that reflects both ancient and modern agriculture. Originating in Germany in the 1920s, biodynamics touts connections between the soil, the plants and trees, and the local animals and insects, as well as using astrological calendars to time work such as planting and harvest.

Ancient rituals still linger in the nearby Caucasus Mountains. Every September Georgians from several ethnic and religious backgrounds—including Muslims—make a pilgrimage to Alaverdi called Alaverdoba. The festival, which now lasts for about a week, was in ancient times a multi-week harvest celebration linked to pagan, pre-Christian moon cults. Anthropologists and church historians have studied and pondered Alaverdoba.

The festival is a touchy subject for the Alaverdi monks, since ritually slaughtering animals is part of the local tradition, and for some years—how long is in dispute—the slaughter was done in or around the church. A 1962 documentary from the Soviet era shows peasants drinking, celebrating, and bringing sheep and other creatures to the festival. One stark black-and-white shot zooms in to a corner of the altar room, showing two bloody and dismembered horned sheep heads on the floor. A man walks in, casually picks them up, and carries them away. The Alaverdi monks imply that such scenes were part of Soviet efforts to discredit the church, but some experts think they are remnants of the ancient moon-cult ceremonies that date back to the dawn of civilization, just like the stone shrines of the Kush nomads, and the cave winery.

Father Gerasim said some of the ruins at Alaverdi were left for a reason. He pointed to a large, low-walled area just outside the main altar room. It was an ancient wine-pressing trough. Archaeologists surveyed the site and dated the remains to the sixth century AD. The trough's surprising size—about twenty by thirty feet—indicates that large quantities of grapes were used in a system that included stone pipes to carry water and waste to and from the monastery. "This is a tradition passing through generations, from the grandparents to the grandchildren," Gerasim said of Georgian winemaking culture. "I remember when I was about three or four years old, my grandfather or father took me to the wine cellar every time they went. Wine ties, and tied, the human being to his community, to his land."

I told Gerasim that I was still trying to understand why grapes and wine are such a fundamental part of Georgian culture, more so than in others. Thousands of farmers still make batches of wine at home each year, using the clay qvevri. He thought for a few seconds and said the location of the old trough provided a clue. The altar room, the oldest part of the church, was built right next to the original winery. Wine wasn't just part of the religious ceremonies. The vineyard was part of the altar. Gerasim didn't say it, but I realized something else. The Georgians had merged far older pagan vineyard rituals into Christianity. The extensive wine and grape symbolism in Christianity and Judaism wasn't a new development, but a continuation of old traditions.

Alaverdi was crumbling not long ago. The Soviets turned the monastery into a truck repair shop, and the monks still find rusty parts buried in the ground. The vineyards of the region weathered challenges, too. Soviet bureaucrats decreed that Georgia's role was to send huge quantities of wine to Russia and Ukraine, but the focus was on cheap factory-produced booze for the masses. Multinational wine conglomerates got interested in Georgia about twenty years ago, after the collapse of the USSR. Consultants scorned the local grapes, suggesting that the most profitable thing would be to plant Chardonnay, Merlot, and other familiar brands. The Alaverdi monks and Georgian government specialists rejected the advice, reasoning that such a strategy would downplay the distinctive aspects of Georgian wine and put the tiny, cash-poor country in direct competition with established wine producers such as France, Chile, and New Zealand.

In the early stages of rebuilding the winery, Gerasim had some doubts. For many years the monks didn't have the money to do all the necessary work, and they often talked amongst themselves, anxious whether visitors from abroad and even the local people would understand what they were making. He paused and smiled. "But all the guests who tried our wine were excited." Over time the local people were, too. "They did recognize that the revival of the winemaking tradition would help them and their own families."

As Gerasim led me around the monastery I asked myself if Georgia's cultural obsession with wine grapes represents the human side of a famous botanical theory. In the 1920s Nikolai Vavilov, a legendary Russian geneticist, theorized that the geographic point of origin of any plant can be identified by finding a region that shows the greatest genetic diversity of that plant. In other words, if you think Georgia or the Caucasus was the birthplace of wine grapes, there should be many, many varieties there because they have had the most time to evolve. And in fact Georgia claims more than five hundred varieties of grapes, with still more in Armenia, eastern Turkey, and northern Iran. So perhaps Georgians are so obsessed with wine because they have such a long history of making it—winemaking began there even before writing existed.

An expedition from the Chicago Botanic Garden and other arboretums found that the Caucasus contains 6,400 distinct groups of plants in a region about the size of Minnesota, compared to 18,743 in the entire United States. The botanists said the interaction of plant communities with existing mountain ranges in the Caucasus help explain the profusion. About 1,600 of the plants in the region—a full 25 percent—are endemic, meaning they are found nowhere else in the world. In layman's terms, the isolated mountain valleys may have helped nurture and protect different species, like natural time capsules.

The region came into existence about twenty-five million years ago, when the African–Arabian and Eurasian landmasses collided. Once separated by a vast sea, they crept closer for hundreds of millions of years, finally buckling upwards to form the Caucasus Mountains. Volcanoes erupted, the ocean shrank, and rivers formed, but in the broadest sense the new region was still a mix of two old landmasses—and all the plants and animals each contained.

That melting pot promoted plant diversity, partly because the geologic parents had both subtropical and temperate regions. As a United Nations report on genetic diversity notes, "This unique situation has made it possible for the Caucasus to be a bridge between eastern and western flora . . . This explains why, in some areas of the Caucasus, species of European or Asian origin grow next to [native] species, adapted to continental, Mediterranean and sub-tropical climates." The field report from the Chicago botanic team went further, adding, "[T]hrough eons of tectonic plates pushing skyward, plants continue to be further isolated—which in some cases has precipitated new species."

Patrick McGovern found Georgian pottery dating to 6000 BC that appears to be decorated with "grape clusters and jubilant stick-figures, with arms raised high, under grape arbors," and burial mounds near Tbilisi contained ornate gold and silver goblets with depictions of drinking ceremonies. "Grapevine cuttings were even encased in silver, accentuating the intricate nodal pattern of the plant," McGovern said. Some of those cuttings are on display at the Georgian State Museum in Tbilisi. Saint Nino, a woman who brought Christianity to Georgia in the fourth century, reputedly carried a cross made of grapevines as a shield against visible and invisible enemies. The classic Georgian chant "*Shen Khar Venakhi* ('Thou Art a Vineyard')" was reputedly written by King Demetrius I in the twelfth century. It is still popular at weddings.

I followed Gerasim to what he called the "new" winery, which dates to 1011. A rambling line of clay qvevri rested against the wall, some of them centuries old, and a few craftsmen still specialize in creating the jars. We walked inside. In the front room were signs of the international support that Alaverdi has attracted. An Italian company donated a wine labeling machine, as well as oak barrels for another product, Georgian brandy. I bent down to pass through a low arch into the qvevri cellar, and Gerasim said the ancient design was intentional, "So people have to bow when they enter the room."

The cellar had old, rough stone walls inset with arched shelf areas, a wood-beamed ceiling, and a tiled floor that was surprisingly bare. There were round holes in the floor above each buried qvevri. Flat stones seal the tops during fermentation, a practical but startling departure from the stainless steel tanks that dominate modern winemaking. "Traditional winemaking in qvevri doesn't need to be updated," Gerasim said. "This is a universal, unchanged wine technology. The only novelty that you can bring is to have some equipment that would simplify your work a little bit. The winemaking itself—no changes." One corner of the wine cellar featured a display that included ceramic drinking vessels and artifacts found on site, as well as an ancient qvevri left buried in place. The monks were rebuilding Alaverdi—and its grapevines—after decades of neglect.

I was excited but a little nervous as Gerasim took me down a stone staircase and into a cool, quiet tasting room. Some of Alaverdi's wines had been honored at major competitions, but would I love them? *Decanter* judges said one red "simply had more depth, more flesh, more length and more spice than its competition."

In the underground tasting room Gerasim took the role of *tamada*, or toastmaster. In Georgia all formal dinners and events have a tamada who gives a series of speeches. Its origins are perhaps linked to prehistoric wine ceremonies, including human sacrifices. The Greek symposiums, where men of rank gathered to drink, tell jokes, and socialize, may have evolved from Georgian traditions. As we prepared to drink Gerasim said, "I can compare our winemaking to a family with many children. The children look like each other, but each of them has their own individual character. They have some things in common taken from their parents, [which are] the ground, the land, and the grapevine. Still, each wine is an individual person by its taste and aroma."

Gerasim declared that guests are a gift from God and that the best food and drink in the house are saved for them. His first pour was the most recent vintage of Rkatsiteli, from 2014. It had a deep golden color unlike any mainstream white wine, the result of leaving the skins on for an extended period during fermentation. Some writers use the term "orange wine" to describe such vintages, but Gerasim said "golden" is the correct term. The wine was beautiful and beguilingly different. It had the crisp, fresh drinkability of a white, yet with deep layers of minerality and flavor. It struck me that this wine solved the age-old question of red *or* white, and that came partly from the qvevri aging process. "They perfectly fit all the dishes at the table," said Gerasim. A bold claim from Gerasim, but I knew that one of New York's leading sommeliers agreed.

Levi Dalton, who has worked with legendary chefs such as Daniel Boulud, told *Wine Enthusiast* magazine that wines produced in qvevri "offer a delicacy of flavors that complement fish, but are structured enough to stand up to a meat course. [. . .] They're like a get-out-of-jail-free card" for food-and-wine pairings. The British wine writer Simon Woolf was impressed, too. He said the Alaverdi Rkatsiteli had "the most astoundingly complex nose of tea leaves, baked apples, jasmine, herbs and plum compote (and bear in mind my description does not remotely do it justice)."

After the first tasting Gerasim poured an older Rkatsiteli. It had even more depth and flavor, yet was still crisp. Then we sampled wine made from Kisi and Khikhvi white grapes. The Kisi had hints of apricot, citrus, and nuts, while the Khikhvi was both flowery and lightly woody. We moved on to wines made from the Saperavi grape, Georgia's leading red variety. It had an inky deep color, with a cascade of flavors: cherry and currant, but also spices and hints of tobacco, with all the crispness of the whites. I was overwhelmed.

As we relaxed in the cool cellar Gerasim recalled once asking his father why he drank a large glass of wine at the end of each day. His father replied, "Work a day in my shoes, manage all I have to manage, and ask me." Now Gerasim understands. I told Gerasim of visiting Cremisan and about all the wineries on my to-see list. "You are a lucky man to visit all these places, and ask these questions," he replied, then said that according to legend Georgian monks brought Caucasus vines to Jerusalem's Monastery of the Cross about a thousand years ago. "It would be quite interesting to conduct some research on that."

I bought a bottle of Alaverdi's Khikhvi to share with friends. Gerasim invited me to come back for one of the qvevri winemaking conferences Alaverdi holds each fall. He surprised me with some kind words: "Very generous wishes that you tell the true story of all the different people making wine from local grapes." I was touched also by his feeling of kinship with other winemakers. There really was a far-flung community of people fighting to preserve rare grapes.

On the drive back to Tbilisi the fields faded away in the dusk. A herd of cows blocked the road in one tiny village; in another an old woman scolded a large black calf for some misbehavior. After so much time talking to scientists and reading histories, the Alaverdi wines were a revelation unlike anything I had ever experienced. Vibrantly robust yet elegant, they were like trying fresh, farm-pressed cider or a raw milk cheese for the first time. The impact of the qvevri suggested what ancient wines may have tasted like: completely unlike wines fermented or aged in stainless steel or oak barrels. In retrospect it made sense. If you cook in an electric oven or that same dish over a fire made with smoky chunks of wood, the food changes character. The qvevri wines are like that.

But it wasn't just the qvevri. Alaverdi was also a testament to how beguiling carefully produced natural wine can be. These wines were unfiltered, made with both the stems and the grape skins left on for the first fermentation. The result was more flavor and depth, not bitterness like you might expect. The tasting was also a reminder that my own preconceptions of ancient wine were too simple, and that much prior commentary had the same flaw. Experts often remark that ancient wine in Greece, Rome, and Egypt contained extra flavorings such as myrrh (a pine resin) or cinnamon to help prevent spoilage. Yet the Alaverdi wine had none of those additives, and it was still wonderful.

The Alaverdi monks believe that wine needs only grapes, wild yeasts, and careful nurturing, and woe to those who imply otherwise. The critic and writer Alice Feiring tells of a German scientist who came to Georgia for a winemaking conference. The scientist commented that native yeasts could cause problems, because bad yeasts might take over the fermentation. One of the Alaverdi monks stood up and demanded, "Are you saying that G-d did not provide the grape with everything it needed to make wine? There are no bad yeasts."

In Tbilisi I saw how Georgia's vibrant grape culture was

attracting winemakers from all over the world. I had dinner with Patrick Honnef, who had left behind a comfortable job in Bordeaux. We ate at a restaurant called PurPur, which is located in a grand nineteenth-century building, with high ceilings and a faded-chic fin de siècle feel.

"I came here first in 2009, and I fell in love with this country very fast," he said. "As a wine enthusiast it's just a beautiful country. Everybody said, you're in Bordeaux, the paradise of wine business. But people don't understand. I'm very happy to be out of Bordeaux. I'm ten times more happy because as a winemaker you can realize yourself far more—you can create here. You can step forward quite fast because the potential of certain varieties is just outstanding."

Honnef loves great French wines, but just like me he sought something different. "As a winemaker in Bordeaux, come on," he said, making a gesture of frustration with his hands. "The only goal there is to match tradition. Stability is the goal, not innovation." Now Honnef is the winemaker for Château Mukhrani, the ancestral sixteenth-century estate of a Georgian prince, located not far from the city. In the 1800s Mukhrani supplied wine to the tsars of the Russian Imperial Court. In 1974 archaeologists discovered the remains of an ancient town nearby, which included a palace mosaic from about AD 150 that depicts a feast of Dionysus, including vines, bunches of grapes, and goblets of wine. Mukhrani was abandoned during the Soviet era, and it wasn't until 2007 that international investors started to rebuild the chateau and its winery.

PurPur combines local dishes with classic French-style cooking, and Honnef ordered a *pkhali* appetizer, which is a sort of pâté made from ground walnuts and different vegetables, such as beets or spinach. The green version looked like pesto; the red version was like nothing I'd ever seen. They both had the taste of eating fresh vegetable shoots straight out of a garden—subtle yet vibrant and alive. I asked about the local grapes as we shared a bottle of Mukhrani's 2013 Goruli Mtsvane, a white that is crisp but full, with peach and citrus flavors. It paired well with the pkhali, and was subtly different than any European wine I'd ever tasted.

Some of the Georgian wines age well, too. A colleague found a case of 2000 Saperavi, and Honnef was impressed. "To taste these wines, for me, was a proof that indeed there's a big potential for aging. It's a pity to drink it very early," he told me. Bordeaux Merlot develops truffle notes after five or six years, while Saperavi gets raspberry-mulberry overtones. Château Murakami also makes some wines that blend local varieties with international ones.

All the positives doesn't mean winemaking in Georgia is easy. "A lot of people [here] did something because their father, their grandfather did it the same way. They don't know why they're doing it," Honnef said of local customs. When he moved to Georgia full-time in 2013, he found that many wineries and grape growers didn't practice basic hygiene, and that despite the quality of the indigenous varieties many wines were "just bad" because of bacteria and related issues. Tradition by itself doesn't guarantee good wine, it is just a starting point.

Honnef's Bordeaux resume didn't carry much weight with rural farmers either. "They're very proud people. I think it's mostly they're afraid of change. But if you show them, and it works, and you show them that it works—you can convince them to try new things," he said.

There are a mix of challenges and benefits. "You have a lot of consultants coming to Georgia, and half of them saying, yeah, you have to plant some international varieties because they're well-known," Honnef said. He gets frustrated sometimes because things don't move fast enough; then reflects on all the country has gone through since the collapse of the Soviet Union. Yet working in a small country has plusses, too. Just before our dinner he met with Georgia's prime minister to review progress in the industry. Despite a modest budget, the National Wine Agency travels to conferences all over the world to give tastings.

Like Father Gerasim, Honnef said he was excited by my plans to explore ancient and forgotten grape varieties. I was, too. Grapevines were everywhere in Tbilisi, literally and figuratively. They grow out of sidewalks, crawl up doors and along balconies, and ramble along working-class courtyards where laundry hangs out to dry. Some people even trained masses of vines into a kind of natural roof for open-air patios. I was used to seeing precisely controlled vineyards. This was different. Georgia was a living textbook. People loved the vines, and I could see it in their art, poems, music, and wine.

But that passion hasn't yet uncovered indisputable proof of the origin of winemaking. Georgians insist their country is the birthplace; neighboring Armenia does, too. Early in my visit I met with Giorgi Tevzadze of the National Wine Agency, which is located in a modest Soviet-era building. Tevzadze told a story that illustrates how difficult preserving rare grapes is, even in a place as wine obsessed as Georgia. During one visit José Vouillamoz found some possibly special vines not far from the wine agency, and he took a sample. "But the evidence that he took, the materials, it unfortunately cannot be traced now," Tevzadze said, because the owner has since destroyed the vines and planted the area with different varieties.

I'd fallen in love with Georgian wines. But would science ever be able to decode the family tree of all five hundred local grape varieties, known or unknown, or their links to Europe? Sean Myles, the Canadian geneticist I had spoken to a while back, explained the challenge. "There was some kind of domestication event over in the East, in the Caucasus, and as the grape was brought over into Western Europe it began to mix with local wild grapes. Wherever the grape ended up, it ended up at least having some contribution from local, wild material," he told me. So German Riesling grapes have more links to their native wild relatives—but they also contain a trace of the Caucasus ancestry.

Part of the barrier to decoding the wine grape family tree comes down to money. "Sure, they might be interested in the origins of Pinot Noir, but they've got bigger fish to fry," Myles said of large corporations, echoing what I'd heard from other scientists. "They're not going to go dish out a bunch of cash for somebody to do some research . . . that doesn't impact their bottom line." I began to wonder if the origins of wine grapes would remain shrouded for the foreseeable future.

Myles brought up something else. As we plant the same grape varieties all over the world and prevent them from cross breeding, that locks in flavor profiles many people love, but also stops evolution. Myles elaborated, "The bad side to that is the pests and things that are attacking those vines continue to evolve. That is going to be the potential demise of the entire international wine industry as we know it today. The industry is losing the arms race to the pathogens that continually evolve and attack the grapevines. It's really only a matter of time. If we just keep using the same genetic material we're doomed."

I'd thought of rare grape varieties from a perfectly selfish point of view. I wanted new flavors and different stories. Myles showed me that diversity is a matter of vineyard health and survival, too. The most notorious example of crop monoculture is the Irish Potato Famine. By the 1800s most people in Ireland had started planting just one potato variety, propagating it from shoots. The potatoes were the same each year, which wasn't a problem until the rot disease *Phytophthora infestans* showed up in the 1840s, destroying entire harvests and leading to massive starvation. I wouldn't wish ill on any grape variety, but that might be what it takes to make industry leaders change their ways.

After I returned from the Caucasus, I took comfort in reading old histories and legends. Many of them do identify the region as the cradle of winemaking. The Greek historian Herodotus described how Armenian traders brought wine to Babylon in the fifth century BC:

> The boats which come down the river to Babylon are circular, and made of skins. The frames, which are of willow, are cut in the country of the Armenians above Assyria, and on these, which serve for hulls, a covering of skins is stretched outside . . . They are then entirely filled with straw, and their cargo is put on board, after which they are suffered to float down the stream. Their chief freight is wine, stored in casks made of the wood of the palm-tree. [. . .] When they reach Babylon, the cargo is landed and offered for sale; after which the men break up their boats, sell the straw and the frames, and . . . set off on their way back to Armenia.

An obscure book added even more color to the Caucasus legends. *Nart Sagas: Ancient Myths and Legends of the Circassians and Abkhazians* is a collection of age-old folktales, including one about the origins of wine. The Narts were a primeval race of demigods who battled giants, witches, and various supernatural forces. They lived in a land of plenty, except for one thing: the giants controlled all the fruits and vines. Obsessed with what they couldn't have, the Narts decided to go to war over the fruits and vines, no matter what the cost. Another Nart saga may be one of the earliest chronicles of men behaving badly while drinking. It is disconcertingly accurate:

> The Old Narts stood over a barrel of white wine.
> Of the harvest god they uttered many blasphemies.
> Aleg, the leader, told many false stories.
> Warzameg, as though true, agreed with them.
> Yimis boasted as was his habit.
> Sawseruquo composed a hundred evil schemes.
> Nart Chadakhstan dreamed of manly deeds.
> In all they drove white wine down to the barrel's seething
>   bottom.

Then Nart Pataraz crashed through the door and crushed a few people's ribs on his way to the center of the room. He launched into a seemingly endless series of boasts about bloodshed and mayhem and laid down a challenge: The mystical barrel of white wine would judge what he said. If it was all lies, the barrel would dry up. If it was truth, the barrel would overflow.

The barrel overflowed, of course.

TASTINGS

You're probably going to have to go all the way to the Caucasus Mountains to taste Alaverdi's wines, which are made in relatively small quantities. Luckily many other Georgian wines are available in the United States, such as Pheasant's Tears. Astor Wines & Spirits in New York City stocks several Georgian wines, and sells them online, too: www.astorwines.com.

Georgian Wine House, an importer based in Maryland, lists numerous stores around the country that stock bottles of Georgian wine: www.georgianwinehouse.com.

If you decide to explore wines from the entire Caucasus region, don't forget Armenia (Zorah Wines, located near the six-thousand-year-old cave site; www.zorahwines.com) and wines from Turkey. VinoRai, a Seattle importer, handles several Turkish wines: vinorai .com/turkey. Look for grape varieties such as Boğazkere (*trans:* "Throat Burner") and Öküzgözü (*trans:* "Bull's Eye").

ALAVERDI MONASTERY CELLAR (Buy whatever year you can find from Alaverdi!)

Mtsvane Kakhuri (amber; made in qvevri)

Rkatsiteli (amber; made in qvevri)

Saperavi (red; made in qvevri)

Kisi (golden; made in qvevri)

CHÂTEAU MUKHRANI

Goruli Mtsvane (white)

Rkatsiteli (white)

Saperavi (red)

Reserve Royale Saperavi (red)

PHEASANT'S TEARS

Rkatsiteli, 2015 (white; made in qvevri)

Saperavi, 2015 (red; made in qvevri)

Mtsvane, 2015 (white; made in qvevri)

Chinuri, 2015 (white)

Tavkveri, 2015 (red; made in qvevri)

ARMENIAN WINES

Zorah Karasì (red, from Areni grapes; made in concrete to lend a qvevri-like quality)

Zorah Voskì (white, from Voskèak and Garandmak [*trans.* "Fatty Tail"] grapes; made in concrete)

8

Yeast, Co-Evolution, and Wasps

God made yeast . . . and loves fermentation just as
dearly as he loves vegetation . . .
—RALPH WALDO EMERSON, "NEW ENGLAND REFORMERS," 1844

The Georgian monk's comment that God doesn't make bad yeast reminded me of another facet of my wine ignorance. I'd never passed judgment on yeast because I never thought much about it at all. I knew that fermentation turns sugars into alcohol and carbon dioxide (thus the fizz in Champagne), and bread-making taught me that wild and commercial strains of yeast are different. But for me and most wine lovers yeast are kind of like bass players in rock bands—essential, but overlooked. So I had never asked the winemaking version of the chicken or the egg question. Are grapes responsible for all the marvelous flavors and aromas in wine—or is yeast? I'd always assumed the former, partly because wine labels, lists, and reviews rarely mention yeast type. Perhaps they should.

In 2016 the Australian Wine Research Institute gave an unusual tasting at a symposium in England. They offered five groups of sparkling wines, with three bottles in each set. At first glance, pretty normal. Each set was made at the same winery, from the same grapes, in the same style. The yeasts were the variable, and they were deliberately bred to produce distinctive flavors.

"It was quite striking how different these sparkling wines were," Erika Szymanski told me in a phone call. She is a full-time scientist and part-time wine writer who works on yeast research. "You could definitely pick up that these yeasts had some distinctive aroma profiles. The room was quite polarized on whether we liked one yeast better or the other yeast better, but I think the universal agreement was, yes they're different, and yes we like the hybrids."

Master of Wine Sally Easton took part in the tasting, and her notes show how much the yeasts influenced the same original grape juice for a white wine:

Wine 4 . . . Smoke and aromatic tar on nose. Savoury, dry bread, hint bitter note, almond maybe, in a more positive frame.

Wine 5 . . . Floral, perfumed, almost muscat nose, repeated on palate. Gentle, lifted style, nice concentration of primary flavours.

Wine 6 . . .Wafty smoke, biscuit and brioche, more classic feel to this wine. Nice balanced dimensions. Wholesome.

One of the specially bred yeasts that Easton, Szymanski, and others liked is known as AWRI 2526. The Australian Wine Research Institute created it by mating a basic version of *S. cerevisiae*, the commercial wine yeast, with *S. mikatae*, a yeast not typically used for alcohol production. The *S. mikatae* was isolated from soil and decaying leaves. The Australians said the experiment created "a new breed of wine yeast" with the ability to increase flavor and aroma complexity, yet still deliver consistent production.

I asked Szymanski my chicken-or-egg yeast question—whether the grape variety or the yeast contributes the most to wine flavors and aromas. She got politely worked up, suggesting that A) it may be impossible to determine; and B) the question is way too simplistic. Her view is that humans and yeast co-evolved.

"When I talk about co-evolution, I'm talking about the ways that humans and yeast have developed closer than usual working relationships. It's not just about a one-direction domestication of something. It's about the way we interact, the way we respond to each other. We have developed long-term relationships that appear to be working well for the humans and the yeast," she told me.

For example, bread yeast and wine yeast seem to have evolved differently. "It's very important to note that of all of the communities of yeast associated with humans, wine strains are pretty robust, they're pretty wild. They have to be very, very robust to deal with all of that alcohol and all of that sugar," she said. But commercial bread yeast don't survive particularly well in environments other than bread dough. That is, they seem to be almost dependent on humans.

Wine yeast have even adapted to our winemaking styles. More than two thousand years ago Egyptian, Greek, and Roman winemakers used sulfur (in the form of smoke) to disinfect clay amphoras. Spanish researchers found that about 50 percent of wine yeast strains contain a genetic mutation that makes them better able to resist sulfur. Wild yeast don't have that mutation. Vineyards have also used copper-based sprays and dust for more than a hundred years. Originally known as "Bordeaux mixture," it was applied in massive amounts to control mildews. As a result, some wine yeast have developed increased resistance to copper poisoning, too.

Yeast use our global wine-lust to travel, too. That's significant because they're not naturally airborne. DNA tests in New Zealand found that while some yeast came from local soils or oak trees, others matched the genome found in French oaks. While insects helped the local yeast spread and find niches in vineyards and wineries, the French varieties hitched a ride on imported barrels.

The travel evidence helps explain another puzzle. Scientists have noticed there isn't much yeast on unripe grapes. Yet when the fruit ripens yeast *is* there—just at the point when grapes have their highest concentration of sugar, which is yeast food. Somehow, the yeast time their arrival. Birds are part of the answer, but some regions are still left with the question of frigid winters, which should kill local yeast off.

Some crafty *S. cerevisiae* found a solution in the guts and life cycles of wasps. Italian researchers captured and dissected nest-building wasps in the spring, summer, and fall. They found 393 yeast strains inside the insects. Some types rose and fell with the seasons, yet *S. cerevisiae* stayed fairly constant all year long. But how? The scientists isolated a small group of female wasps in the lab, inoculated them with a specific strain of *S. cerevisiae*, and let them hibernate. Three months later that yeast was still alive and well. The research showed that wasp queens can transmit yeast to young larvae and workers during the next life cycle. In the spring and summer they chew up food (mostly insects), regurgitate it, and feed their young. In other words, "wasps can maintain a potentially unending transmission of yeast strains through favorable and unfavorable seasons . . ."

The scientists don't claim that wasp bellies are the only way yeast survive winter—they can live inside beehives, too. But they found that yeast strains in wasps and grapes from the same vineyard are very similar, even in different seasons and years. That suggests something I'd never suspected: not only does yeast play a role in creating distinctive regional wine flavors and aromas, wasps and other yeast-hosts are partners in the effort, too.

Other yeast research supports some theories McGovern and Vouillamoz have about the origins of winemaking. French researchers examined yeast DNA and found that most wine strains had a common parent region in ancient Mesopotamia—the Fertile Crescent. The wine yeasts spread across the Mediterranean and up the Danube River Valley, just like winemaking. The studies indicated that wine yeast and humans have had an intimate association for thousands of years. They also found that bread, beer, wine, and sake yeast diverged into groups—tribes, in a way—about ten to twelve thousand years ago. Each different yeast found a niche in a human industry, and mostly stayed there. The genus name, *Saccharomyces*, derives from the Greek words for sugar and fungus, and the species name, *cerevisiae*, comes from a Latin word for beer. Some ancient yeast DNA was also very similar genetically to strains McGovern isolated from Egyptian amphora.

Another team of researchers looked at modern yeast DNA, expecting to find a neat genetic trail leading back to a common ancestor. Hardly. It turned out that humans weren't the only ones taking part in wine-fueled ancient orgies. The yeast cells hadn't just split in two, which is how they normally reproduce. Perhaps helped by all the unwashed feet that smooshed grapes over the millennia, some had hooked up with different yeast species. So even within the general tribes of wine, beer, or bread yeast there's some variety.

There's still disagreement over where yeasts originated, and when. In the 1990s Raúl Cano, a professor emeritus at California Polytechnic State University, recovered S. *cerevisiae* DNA from the guts of 25-to-40-million-year-old bees and termites trapped in amber, and later from a Lebanese weevil that may have lived 125 million years ago. Fruits and other organic materials fermented long before humans showed up, so clearly yeast existed before wine. But for thousands of years no one really understood what yeast was, beyond some bubbling force that makes alcohol, bread, and fermented food such as pickles and kimchee.

By the early 1800s biologists began to realize that yeasts are separate, living organisms, which suggested that fermentation was more than a type of rot. Louis Pasteur began doing serious fermentation research in the 1850s, and he is generally credited with discovering the full role of yeast. Pasteur later wrote *Études sur le vin*, a four-hundred-plus-page survey of winemaking and vineyard diseases. It was the dawn of serious wine science.

Yet the idea of replacing wild wine yeasts with a commercially produced strain didn't emerge until the 1890s, and that research languished for decades. Winemakers felt the tradition of using wild yeasts had worked for thousands of years, so why change? But change did come, in 1965, when a California winery used two dried yeast strains manufactured by Red Star Yeast. By the eighties many large wineries used commercial yeast for some or all of their winemaking. At first glance, so what?

All the research suggests there was once a huge diversity among wild yeasts and even regional wine grape yeasts, but there's a problem. Commercial wine yeasts are taking over, literally driving some of the wild populations towards extinction. Researchers sampled yeast populations in a new Spanish winery for five years. At first local non–*S. cerevisiae* strains dominated the first stage of fermentations. But over time the commercial yeast pushed out the local ones. Other researchers found that while there are hundreds of different commercial wine yeast products, many of them are virtually identical genetically.

I saw parallels to *Wine Grapes*, though I doubt there's the same market for a thousand-plus-page companion called *Wine Yeasts* (though one never knows). In any case, yeast clearly affect wine flavor, and their diversity is threatened, too.

A quirky twist left me hopeful and puzzled. Cano, the scientist who extracted the multimillion-year-old yeast from amber, tried to patent the primordial organisms. He originally thought they would prove useful or profitable in biotechnology or medicine. That didn't work out, but he cultured some of the yeast spores and realized they could make a beer.

In 2008 Cano started offering trial batches of Fossil Fuel brew, and in 2016 he continued the experiment with Schubros Brewery. An *Oakland Tribune* critic said the yeast gave the wheat beer a distinctively "clove-y" taste and a "weird spiciness at the finish," while the *Washington Post* deemed it "smooth and spicy, excellent with chicken strips." After the brief burst of publicity, though, the project seemed to disappear. Schubros brewer Ian Schuster told the *San Francisco Gate* the yeast was unpredictable and "high maintenance," prone to delivering different tastes at different temperatures. "It needs to be roused," he added somewhat mysteriously.

I found some possible explanations in other scientific papers, which suggested that the primordial DNA was impure, or degraded. Or perhaps it was a glimpse of the distant past when even yeast was "finding its way," so to speak, learning how to best transform sugars into alcohol and carbon dioxide. As a business venture the ancient yeast may just be too finicky to deliver any consistent flavor, which is a big problem.

The Fossil Fuels experiment illustrates a bigger point. There are still a lot of questions about how yeasts do their fermentation magic. Clayton Cone, a scientist with Lallemand, a French company that specializes in food yeasts, said in a recent article that even he has "only a rudimentary understanding of what is going on inside the yeast cell." What's clear, however, is that over millions of years countless yeast species adapted to highly specific environments—including within grapes, wine barrels, or other winery equipment.

The US Food and Drug Administration approved a genetically modified (GM) wine yeast in 2003. So far it hasn't led to huge controversy, perhaps because the FDA doesn't require winemakers to tell consumers whether it's used, since yeast are technically processing agents and not ingredients. The GM strain is now marketed by Lesaffre, a French food science company with roots that go back to 1853.

The GM yeast was approved for use in the United States, Canada, and South Africa, so many people have probably already drunk wines made with it. A closer look at the new yeast shows how alluring such products might be for producers. The technical details are that it allows so-called secondary, or malolactic, fermentation to take place during the traditional first fermentation, saving time and perhaps making quality control easier for the winemaker. The yeast's creator said it can eliminate by-products in wine that produce off-flavors, and perhaps even those which lead to headaches or other negative health effects. (There's no proof of those last claims, by the way.)

Szymanski wrote in a blog post that humans may not make the best long-term choices for the yeasts or for flavors. "[O]nly yeast with mutations that do something we like will be picked up and propagated *en masse* instead of being weeded out as undesirable. This probably doesn't bother you. Unless you, like me, stay up at night worrying about microbial genetic diversity . . . But what if we're letting the strongest persuade us to act in their favor at the expense of listening to all of the voices involved or considering long-term environmental policy?"

In other words, what if the oddball yeast play important roles in vineyard ecosystems? That's another way of suggesting the Georgian monk was right. In the big picture there are no bad yeasts.

9

Aphrodite, Women, and Wine

I looked out at the Mediterranean on an early summer day and speculated, as many have, why Homer once described it as "wine-dark" in color. Did some wine have a dark blue or dark green tint, or was it simply a poetic phrase for how the sea looks under stormy, rain-swollen clouds? Experts have suggested all sorts of possibilities, including color blindness among ancient Greeks.

As I walked around the ruins of a temple to Aphrodite on the island of Cyprus, I could imagine slave-powered galleys on the horizon. Crowds once gossiped here at the public baths, perhaps drinking wine and looking out at the sea just two hundred yards away. Most of the temple columns were creamy-yellow limestone, but a few had dark, dark marble laced with natural white swirls and embellished with carved spiral flutes, making them look alive with movement.

Cyprus was one of the earliest places winemaking appeared in the Mediterranean, after spreading from Egypt and the Levant. Just fifty miles south of the Turkish coast and a hundred miles west of Lebanon, the island was strategically located on early trade routes. Cypriots showed a talent for buying, selling, and making wine, perhaps as far back as five thousand years ago. The pharaohs of ancient Egypt came there to pick up perfume, copper, and wine; so did the Phoenicians, the Greeks, the Persians, the Crusaders, the Ottomans, and the merchants of Venice. They all drank, bought, and fought over Cypriot wine. At the time the nearby city was called Amathus; the ruins are a few miles outside Limassol, on the southern side of the island.

The Cyprus visit also made me curious about women's role in the origins of winemaking. The ancient Greeks believed that Aphrodite ("Venus" to the Romans) was born in the Cyprus surf, and she's linked to early female vine goddesses. At the temple ruins I sat down in front of two grottos, called *nymphaeum*, nestled against the hillside at the edge of the ruins. About six feet wide, twelve feet long, and four feet deep, with an arched alcove at one end where statues once stood, the sanctuaries honored nymphs and goddesses, and were probably covered with flowers, ferns, and symbols of nature.

The earliest nymphaeum were often near natural springs; they later appeared in palaces, temples, and public baths all over the ancient world. Marriages sometimes took place at nymphaeum, along with cult rituals that held on until the dawn of Christianity. More than anything they were women's territory. It seemed fitting that one of the few truly female wine grapes in the world is still planted on several dozen acres in Cyprus. Unlike almost all vineyard grapes, Maratheftiko, a red, isn't a true hermaphrodite. It can't self-pollinate. I was curious how and why farmers kept it, and what the wine tastes like.

The sun began to scorch, so I left the ruins to meet sommelier Georgios Hadjistylianou, my guide to Cypriot wine. The island's glorious reputation for winemaking had faded by the 1980s. Native grapes suffered amidst the same tide of French wine that engulfed much of the world. About twenty years ago, though, a handful of vineyards started focusing on local varieties. Was great Cypriot wine returning, or was it like Aphrodite's ruins—a colorful but broken echo of the past?

Georgios is that rare local wine booster who combines love of country with international perspective. Cyprus-born, he's of average height and weight, but with an outsized personality. He has "In Vino Veritas" tattooed on his left forearm and "Riesling" and "Assyrtiko" on the right, and he is a classic sommelier—passionate about wine and food, but willing to deliver stern judgments when necessary. Georgios worked in New York City restaurants for many years, including the famous Monkey Bar on East Fifty-Fourth Street; he returned home in 2008 as co-owner of Fat Fish, an airy Limassol restaurant right on the beach, next to a yacht club.

Over several days we sampled Cypriot wines made from grapes I'd never heard of: Maratheftiko and Spourtiko reds, and Xynisteri whites. We ate *halloumi*, a robust, rustic goat and sheep milk cheese that can be grilled like a steak, as well as fresh squid and tiny, lightly fried fish. The local wines went perfectly with the food, yet Georgios told me that not long ago many people here shunned native grapes. This despite thousands of years of praise for Cypriot wines, including the semisweet red Commandaria, often considered one of the oldest named wines in the world. The Greeks and Romans loved it, and the English king Richard the Lionheart served something like Commandaria at his wedding in 1191. In the medieval French poem "The Battle of the Wines" Cyprus took the top spot in a contest seeking the world's best wine. In the 1600s Italian playwrights praised Commandaria and a Cypriot white.

In 1970 the Cypriot Ministry of Agriculture started promoting popular French grapes as part of a replanting program. Georgios told me that by the 1980s it was more fashionable to drink French wines. People suggested that local grapes such as Xynisteri didn't age well, partly because few restaurants of the era had proper wine storage facilities and poor storage led to oxidation. The big wineries chased easy money, selling huge batches of cheap alcohol to Russia and other Eastern countries.

Somehow, I'd hoped that a small Mediterranean island with grand wine traditions had escaped such forces, but no. Georgios said a few country people clung to homemade wines aged in clay containers half-buried in the ground, much as the Georgians do. They also use the entire grape cluster, including the stems, another holdover from ancient winemaking.

I asked Georgios if history inspired him, since Cypriots had been making wine for at least five thousand years. "Absolutely," he replied. "Look, we were one of the first eight or ten nations to produce wine. How come the French, the Italians, the Germans, the Austrians—they're way ahead in the game? I think very few people [here] had the knowledge. We produced wine, but it was more home consumption. I think this generation of producers, which started mid- to late-nineties, made great strides. The industry has changed tremendously. Now people are more knowledgeable." As local wineries refocused on native grapes and improved vineyard management, customers responded. "We'll see better, hopefully with the next generation, with their kids, or whoever is going to continue their work."

"So now they're keeping the local grapes, but using international standards of fermentation, refrigeration, and product handling?" I asked.

"Correct," he said, and asked if I wanted to try some. I certainly did. He poured a Xynisteri from Tsiakkas Winery. It tasted of honey and flowers, with a nice balance. Georgios said it becomes drier as it ages, with hints of herbs. I asked how a tiny country could re-introduce such wines to the world, noting that there was no local Robert Parker promoting native grapes. "Parker did exceptional things for wines," Georgios said of the man who became famous—and infamous—for grading wines on a hundred-point scale. "I wish we had Parker here. I think he sometimes gets blamed, but it's not his fault that so many winemakers try to please him. It's like if there's a pretty girl and all the guys are chasing her."

Georgios hopes that as Cypriot wines keep improving, they'll become better known. "The first time we went to New York with the Greek varieties [in the 1990s] everybody wanted to taste them, and be the first one that 'discovered' Assyrtiko," he said. Fat Fish recently built a temperature-controlled wine cellar that can hold five thousand bottles, so they'll be able to age local wines. I wanted to meet one of this new generation of winemakers, and Georgios mentioned Marcos Zambartas.

The Zambartas Winery is located in a tiny village about fifteen miles outside of Limassol, in the foothills of the Troodos Mountains, which are covered with cypress and pine trees and snow for several months of the year. Byzantine monasteries dot the region, some dating back to the eleventh century. Copper mining began nearby around 3000 BC. Both the tree and the island are probably named after Cyparissus, a mythical boy who accidentally killed his pet stag and became so grief stricken that he turned into a tree.

Marcos Zambartas is a handsome, friendly man who studied winemaking in Australia. He met his wife there, and after training in New Zealand and France the couple came back home in 2007. Marco partnered with his father, who had cataloged native grapes decades earlier, when almost no one cared. Their winery is built into the side of a hill, with a living area upstairs and the fermentation, aging, and tasting rooms below, so the earth keeps the wine cool year-round. Zambartas, who has a chemistry degree, wanted to apply international winemaking skills to the local grapes, and change a culture that had floundered. "Until twenty years ago the focus was just on turning juice into alcohol. There was not a real understanding of the indigenous grape varieties. Whatever happened, happened," he told me.

His late father, Akis Zambartas, set out to find native grapes in the late 1980s, seeking the forgotten vines that elderly people said their grandfathers had planted. He found twelve varieties over a three-year period. Some had names. Many did not. He kept those scarce, unknown vines because of their colors, their aromas, and their tastes. A local priest offered some land for a rare grape nursery. Akis replanted vines and tested micro-vintages made from grapes such as Promara, Spourtiko, Flouriko, Yiannoudi, Kanella, and Omoio. Experiments with Maratheftiko showed a possible link to early wine grape domestication; though it doesn't self-pollinate well, some long-ago farmer learned to plant Spourtiko grapes next to every third vine, to boost pollination. That's still done on Cyprus.

Today Zambartas is experimenting by using the native grapes with various French styles of winemaking. His 2014 Xynisteri used grapes harvested from a single vineyard, aged in a combination of oak and stainless steel. It was a beautiful white wine, very crisp but full-bodied and spicy. Subsequent vintages had hints of cinnamon and honey. "Slowly, slowly, we are limiting the contribution of the international grape varieties," he said.

We tried his 2013 Maratheftiko, a silky, full-bodied red with violet and cherry aromas, and the potential to age beautifully. It seemed like madness for locals and tourists to ignore a wine like this in favor of Burgundy or Bordeaux knockoffs. Yet some travel hundreds or thousands of miles to see the ancient Cypriot churches and archaeological sites, eat the local seafood and cheeses—and then order imported Chardonnay or Pinot off the wine list.

Zambartas showed me the winery, where my questions weren't just about the past. I asked if his chemistry degree had led him to winemaking. He said it helped. "One of the reasons I chose winemaking is because it's the most fascinating application of chemistry." And it's true—winemaking is about the grapes, soil, fermentation, and aging all coming together to create flavors. "The mechanisms, they are understood to some extent, but we don't have the full picture yet. It's very complex chemistry in a sense [and] it just shows how little we know at the end of the day. Of course the aim of the plant is to attract predators [with flavors] to get the seeds and propagate the species."

How flavors emerge and change chemically as wine ages flummoxed me. Zambartas knew and described how tannins soften out. "It is unpredictable, but the science behind it, it's understood. Tannins at the beginning, they're smaller molecules, and with time, they bind together at specific points of the molecule." As they transform, aromas such as leather and cigar box emerge. "These aromas are because of the aging in the bottle. It's not because of the barrel, it's not because of the fruit, it's not because of the fermentation," he said.

When tannin molecules get too big they fall to the bottom of a bottle. That's why wines lose color over time, since the tannins often contain wine pigments (such as red). Zambartas said oxygen acts to bind molecules, but it is fickle. Tannins normally absorb oxygen as part of the aging process, but if the wine absorbs oxygen too fast, the tannin molecules can't keep up. Then aromatic molecules start absorbing the excess oxygen, which degrades them. That's why wine needs to be stored at around 60 degrees Fahrenheit. Once it gets much warmer, the molecules and thus the aromas change.

The trickiest chemistry is balancing tradition with innovation. Zambartas wants to modernize the taste of Commandaria— roughly the Cypriot equivalent of changing the Coca-Cola formula. Customers told him the traditional Commandaria was too high in alcohol, and too sweet.

"We made one in 2011, that was the first year. We made another in 2012, it's still [aging] in barrels," he said. "So our philosophy . . . is to not fortify it." Instead of the usual 15 percent alcohol, his is 12.5 to 13 percent, with higher acidity. "We haven't bottled yet, but with a limited number of tastings, we've gotten very good feedback. We need to reinvent it without changing the product completely, which would be a sacrilege." I tried a sample of his Commandaria from the barrel, as well as a popular brand from a larger winery. There was no comparison—I liked Zambartas's far more.

We left the cool cellar, walked outside, and looked across the small valley. I saw a few vineyards nearby, but not many, despite all the stone terraces. "This used to be vineyards, this mountain, and now almost none of it is," he said sadly. "In the summer the mountains used to be green, and now they're brown. It is a memory of the past, all these stone walls."

Yet Zambartas is optimistic. More and more people are noticing his wines. A small group of visitors from the British Cayman Islands, attracted by the positive local reviews, arrived as I was leaving. They loved the tastings. Zambartas admits that it was somewhat crazy to invest in a winery on Cyprus, but he and his wife are building connections in the international community. He partnered with Angela Muir, a British Master of Wine, on a program that brings winemakers from other countries to work in Cyprus. Men and women from Australia, France, and New Zealand have taken part. Visitors can also volunteer during harvest time, and receive some free bottles in return.

Before leaving Cyprus I visited another winemaker. During a tour of a newly planted vineyard I asked if the surrounding country people welcomed such efforts. No, he said, they wait for you to fall on your face with any project involving new ideas—even the replanting of local grape varieties. It was an echo of the resistance to change I heard about in Georgia and Israel, a reminder of how fragile grand traditions can be. Cypriot wines had an outstanding reputation throughout Europe for thousands of years, yet were almost pushed into oblivion in favor of cheap plonk and a few French varieties that aren't suited to the climate.

There was more perspective at the small Cyprus Wine Museum, which opened in 2004. There were fine archaeological and cultural displays that suggested winemaking began on the island about five thousand years ago. That estimate fit with outside research, and scientists have also begun to show where the Cypriot vines fit on the overall family tree of wine. DNA analysis revealed that the Cypriot Malaga grape is genetically similar to Muscat of Alexandria, so it was probably imported from Egypt thousands of years ago. Other evidence suggested more convoluted ties. The Cypriot Moscato and Bulgarian Tamyanka are really the same grape; and they're both related to Greek Moscato Kerkyras and Italian Moscato Bianco. No one knows which country planted them first, but it probably wasn't Italy, given how winemaking spread from East to West. The DNA of the Cypriot grape Siderits is even more perplexing, since it isn't closely related to other local varieties, or to regional ones. It could be a long-lost relative of a wild vine that grew on the island millennia ago, a variety introduced by one seafaring trader, or even one that evolved from a seed a bird carried to the island in the distant past. I was beginning to see that José Vouillamoz's dream of mapping the entire wine grape family tree was far more complicated than I had imagined.

A stop in neighboring Greece showed what may be possible in Cyprus. In the 1990s the government launched agricultural and marketing programs to support local wines, which are now widely available in restaurants and wine stores across the United States.

Greece has so many wine-producing regions that I felt as though I cheated a little by stopping briefly in Athens. At the cozy wine bar Heteroclito, co-owner Marie-Madeleine Lorantos offers a staggering diversity of wine by the glass and bottle. Assyrtiko is the country's classic white, minerally and super-crisp. Sigalas is a big, fruity, buttery red, and Avgoustiatis has beautiful, soft, violet aromas. I tried a glass of Xinomavro red. It was fruity and spicy with a distinct tomato aroma. I'd never tasted anything remotely like it.

I got a sense of Athens's past by walking up the looping paths to the Parthenon. Completed around 438 BC, pictures don't do full justice to its massive scale: 228 feet long by 101 feet wide, and nearly 50 feet high. The Temple of Olympian Zeus, on lower ground about one mile away, was even larger. I loitered in both places on one afternoon and stopped in front of a statue from the Theatre of Dionysus, the god of wine, where the great Greek tragedies and comedies were performed before crowds of up to sixteen thousand. The statue was of Papposilenos, a half-man-half-beast and a friend of Dionysus who was reputedly drunk all the time. The sculptor captured the sense of a wild-eyed, massive creature, with flowing beard and hair and a wooly torso, like the hide of an animal. There was a yearly Dionysia festival, held in two parts, one rural and one urban. Dionysia were characterized by Plato and others as "bacchic revels or orgies of women in honor of Dionysus, [which] carried away the participants despite and beyond themselves." A play by Euripides has the following chorus:

> Blessed is he who, being fortunate and knowing the rites of the gods, keeps his life pure and has his soul initiated into the Bacchic revels, dancing in inspired frenzy over the mountains with holy purifications, and who, revering the mysteries of great mother Kybele, brandishing the [staff], garlanded with ivy, serves Dionysus.

After returning home from Greece and Cyprus I continued reading about Aphrodite and the Dionysia. As winemaking knowledge moved west across the Mediterranean I could sense parallel changes: religions and cultures were ever so slowly moving away from the pagan rituals of the Caucasus Mountains and the Fertile Crescent.

I was used to modern versions: seeing Aphrodite as the seductive, almost innocent young woman of the Venus de Milo at the Louvre, or in Botticelli's famous painting *The Birth of Venus*, which shows her perched in a scallop shell with flowing hair—an allusion to her Cypriot origins in the sea. But the early versions of her myth tell a more complicated story. Followers on Cyprus also worshipped a bisexual version named Aphroditus, later Hermaphroditus, who had a beard, breasts, and male genitals, and Cypriot seafarers worshipped a female warrior version of Aphrodite. There were Aphroditus rituals related to wine, marriage, and sacrifice, all fleeting remnants from the earliest cultures of the Fertile Crescent.

Patrick McGovern found that both human and divine females played key roles in making and selling fermented beverages in Mesopotamia, Egypt, and Anatolia. Azag-Bau, a female wineshop owner, rose to high levels of society in the Egyptian Kish dynasty of 2400 BC. Women were clearly some of the earliest winemakers, entrusted to make (and sell) a precious, powerful commodity. James George Frazer, in his landmark study of myth and religion, *The Golden Bough*, wrote that Cypriot ceremonies for Aphrodite closely resembled the Egyptian worship of Osiris, and her cult was perhaps linked to the even older Hittite goddess Ishtar, who tried and failed to woo Gilgamesh.

The Babylonians had revered a goddess named Geštinanna, the mother of all grapevines. Siduri is a major character in *The Epic of Gilgamesh*, regarded as the oldest literary poem. She's like a Mesopotamian version of Adriana La Cerva from the *Sopranos*; a winemaker, tavern keeper, sex symbol, and confidante to wandering men. Alarmed by Gilgamesh's violent, ragged looks, Siduri bars the doors when he first approaches her tavern. She eventually lets him in, tries to dissuade him from a perilous journey, and after listening to his boasts gives perhaps the first example of carpe diem ("seize the day") advice—plus tips on how to behave around women and children.

> But you, Gilgamesh, let your belly be full,
> Enjoy yourself always by day and by night!
> Make merry each day,
> Dance and play all night!
> Let your clothes be clean,
> Let your head be washed, may you bathe in water!
> Gaze on the child who holds your hand,
> Let your wife enjoy your repeated embrace.

Greeks and Romans allowed occasional wild celebrations that sound like an early version of girls' night out. Cults dedicated to the wine god Dionysus (and the later Roman version, Bacchus), set aside three days of the year for all-female celebrations. "[It] seems evident that participation in the Dionysian *orgia* afforded Greek women a means of expressing their hostility and frustration at the male-dominated society, by temporarily abandoning their homes and household responsibilities and engaging in somewhat outrageous activities," historian Ross Kraemer wrote.

Some ceremonies merely involved dancing, snake handling, or walking in the mountains at night, while others let women explore all types of sexual freedom. Men were sometimes allowed—but only when they dressed as women. The Greek Philostratus described a ceremony in the third century AD wherein "torches give a faint light, enough for the revellers to see what is close in front of them, but not enough for us to see them. Peals of laughter rise, and women rush along with men, [wearing men's] sandals and garments girt in strange fashion; for the revel permits women to masquerade as men, and men to 'put on women's garb' and to ape the walk of women."

As I learned more about Aphrodite's origins, ancient Cyprus emerged as both a literal and metaphorical turning point in the story of wine. Three, four, or five thousand years ago, Eastern celebrations boldly mixed wine, women, sexuality, and fertility. Greek, Roman, and eventually European civilizations all kept the wine aspect, but over time began to limit or frown on the public pagan rituals.

Macrobius, a Roman who lived in the fifth century AD, describes polysexual rituals in a book about the Saturnalia holiday (which honored the god Saturn):

There's also a statue of Venus on Cyprus, that's bearded, shaped and dressed like a woman, with scepter and male genitals, and they conceive her as both male and female. Aristophanes calls her *Aphroditos*, and Laevius says:
    Worshipping, then, the nurturing god Venus, whether she is male or female, just as the Night-shiner [moon] is a nurturing goddess.

A double standard eventually developed, however. Greek and Roman men could openly drink, but sometimes women risked harsh punishments. As early as the seventh century BC Romulus said wives who drank could be sentenced to death, and around AD 30 Valerius Maximus reported that one man "beat his wife to death because she had drunk some wine. Not only did no one charge him with a crime, but no one even blamed him." Other writers lamented that women still drank, even though it was sometimes in secret or for medicinal reasons. Juvenal, a Roman satirist who lived in the second century AD, seemed outraged and at times afraid of the passions that wine released in women.

"What decency does Venus observe when she is drunken? when she knows not one member from another, eats giant oysters at midnight, pours foaming unguents into her unmixed Falernian, and drinks out of perfume-bowls, while the roof spins dizzily round, the table dances, and every light shows double!," Juvenal wrote in one passage. He later complained that even sacred rituals were being debauched: "[The] mysteries of the special Goddess of Women [Bona Dea] are no longer secret! Women get all stirred up with wine and wild music; they drive themselves crazy; they shriek and writhe—worshippers of Phallus. And sex. [. . .] The temple rings with the cry, 'Bring on the Men.' Soon they need replacements; when they run out, they jump the servants . . . We can't even lock the women up to keep them in check. Who'd guard the guards?"

Women fought back against all the judgmental and controlling mansplainers, too—at least on stage. In 400 BC the playwright Euripides wrote *The Bacchae* about women who fall under Dionysus's spell. In one passage the princess Agave and a group of women who are peacefully resting in the forest go on a rampage after men disturb them. "[M]en are hunting us down! Follow, follow me! Use your wands for weapons," Agave cried, and the men fled. Dionysus relates what happened next:

> [We] barely missed being torn to pieces by the women. Unarmed, they swooped down upon the herds of cattle grazing there on the green of the meadow. And then you could have seen a single woman with bare hands tear a fat calf, still bellowing with fright, in two, while others clawed the heifers to pieces . . . Then the villagers, furious at what the women did, took to arms. And there, sire, was something terrible to see. For the men's spears were pointed and sharp, and yet drew no blood, whereas the wands the women drew inflicted wounds. And then the men ran, routed by women! Some god, I say, was with them.

Historians haven't found evidence that ancient women really went on rampages while worshipping Aphrodite and Dionysus, but the Euripides play clearly reflected at least subliminal tensions.

TASTINGS

It's easy to find Greek wines in many stores and restaurants across the United States, and online. Cyprus is a bit harder—you might have to go there! I strongly suspect some of the Cyprus reds will age well, but that's just a guess.

CYPRUS

Zambartas Winery

Other small, respected Cyprus wineries include Ezousa, Tsiakkas, and Vlassides.

GREECE

Domaine Glinavos, in the village of Zitsa

Methymnaeos Organic Wines, on the island of Lesvos

Gaia Wines, Athens

"Ritinitis Nobilis" Retsina Estate Argyros, on the island of Santorini

Domaine Foivos, on the Ionian Islands

Papia, western Macedonia Domaine Sigalas, on the island of Santorini

Mavrotragano–Mandilaria (red)

PART TWO

Wine comes in at the mouth
And love comes in at the eye;
That's all we shall know for truth
Before we grow old and die.
I lift the glass to my mouth,
I look at you, and I sigh.

—WILLIAM BUTLER YEATS,
"A DRINKING SONG," 1910

PART TWO

10

Goliath, Foraging, and an Answer

Hide our ignorance as we will,

an evening of wine soon reveals it.

—HERACLITUS, CA. 500 BC

I arrived in Israel again just as a colossal, early September sandstorm smothered the entire region. Flights were cancelled, children stayed home from school, paramedics ministered to hundreds of people in respiratory distress, and Jerusalem reported air pollution levels 173 times higher than normal. I wanted see if Cremisan was surviving, struggling, or thriving. Deep down I also hoped for one more chance at understanding the red Cremisan wine that had captivated me in the hotel room years before. Now I loved their white wines, but the new red didn't have any of the spiciness or earthiness I remembered. Something had changed, beyond just the winemaker.

I also planned to go on a foraging expedition with food historian Uri Mayer-Chissick, whom I'd met in Israel the previous spring, and tour the archaeological dig at Gath, Goliath's home-town, with Aren Maeir, the archeologist I had met on Emek Refaim Street. Which was perfect, because I still wanted more of a feel for what ancient life was like in the region.

For a few days the sandstorm had me worried. The haze of biblical proportions was relentless, but it finally blew over. Mayer-Chissick gave the green light, and said the storm told a lesson about the past, too. People could know how and when to forage only with detailed local knowledge: which plants were tasty, when and where they were abundant, and even names for all of that in order to communicate the risks and rewards to other members of a tribe or culture. "You need to see what's around you. If it's too hot, if you have a dust storm, you change plans," he said. The same was true for ancient winemaking, I thought.

Mayer-Chissick's wife, Tali, inspired his food career. Years earlier he started cooking for her; then he wanted to know more about where food comes from and what people ate throughout history. His academic and personal interests have grown into a family business. The couple now offers a variety of programs and events that promote healthy eating and healthy communities. He believes solutions can come from Jews, Arabs, and Christians, together, since they share many of the same foods.

We were headed to the Biriya Forest, a national park in Upper Galilee, but there was a local bounty at the kibbutz, too. Date palms were loaded with bunches of fruit, free for the taking. I eagerly filled a bag with two varieties. They were like fruit-honey bombs, exploding with flavor.

Most Jewish holidays are connected with agricultural traditions, such as the planting season or the fall harvest, Mayer-Chissick said, and many Mishnayot (religious common law) refer to specific ways to handle food or wine. That kind of knowledge helped people survive in ancient times. I mentioned that some experts claimed that ancient wine was all bad, partly because of spoilage and contamination. Not true, he said. Ancient people spent a lot of time figuring out how to store and preserve food. The soft limestone of the region was perfect for creating underground cellars, cool in summer and warm in winter. Jews also had specific guidelines for how to handle wine, since grapes are considered one of the original seven biblical foods mentioned in Deuteronomy.

We arrived at the forest before the others who had signed up for the trip. Mayer-Chissick immediately set off to look for food, and I followed. One grove of figs had lost most of their fruit, but some had dried on the tree, leaving chewy pieces to snack on. A wild fennel plant was nearby, and we plucked off pieces. A carob tree still had tasty pods, despite the heat.

Sharoni Shafir, a researcher of bees at The Hebrew University of Jerusalem, arrived for the tour with his wife and three children. He lamented that local food traditions were in danger of vanishing in Israel, so it's important to conserve not only biological diversity but also genetic and cultural diversity. Shafir wanted his children to see the bounty that nature provided, and to know how people of the region used to gather food, before such skills are forgotten.

When the whole group was assembled, we trooped around the forest. Children climbed up walnut trees to pluck nuts, and used stones to break them open. We found wild capers, pistachio trees, Palestinian buckthorn berries, sumac seeds, pine nuts, and a few fruits that weren't ripe yet, such as spicy hawthorn berries and olives. Mayer-Chissick combined the haul—mostly spices and flavorings in this case—with lunch cooked over an open fire.

Everyone left happy, but on the ride back to the kibbutz Mayer-Chissick and I commiserated about the challenges of preserving local foods or wines. "We're ruining the traditions. We're losing all the knowledge that was gathered in thousands of years. In the last few years people are talking more about healthy food, local food. But it's very, very slow in Israel. We have a lot of work to do," he said.

That helped explain why at one point Cremisan was the only winery using local grapes. A longtime resident of the kibbutz later provided more insight. Until the 1970s, Israel was still a relatively poor country, worried more about survival than wine. Many of the people who helped found the country after World War II had fled from Europe, so they were used to those wines, not strange Middle Eastern grapes. As the country prospered, young people drifted away from the rural kibbutz lifestyle. Modern, urban life was more attractive, and frankly, easier. It reminded me of what José Vouillamoz said about the younger Swiss, that they didn't want to work in the fields or the vineyards.

Mayer-Chissick is trying to rekindle awareness of rural life with all sorts of different projects. There is a plan to renovate the ancient market district in the city of Lod, which has been plagued by poverty and violence. An environmental park is in the works just a few miles away in Jordan. "We need to talk to them. Cooperate. Get to know them," he said of Jordanians. "I think that's a very, very good way to connect with neighbors. The same land grows the same fruits and vegetables, and the land has the traditions. I think that the local traditions, they don't really belong to the people, they belong to the land."

I felt the same way about native grapes.

THE LAST THING I expected to see outside Cremisan's winery was a tour bus, but there it was. Thirty or forty Brits streamed out, pale as if they were in London. Three Norwegians arrived from Bethlehem by cab. Winemaker Laith Kokaly and agronomist Fadi Batarseh, both Palestinians, greeted everyone in the main bottling room. It was all encouragingly normal. Cremisan was behaving like a business. A dignified, silver-haired man said he greatly enjoys the wine, and was surprised to find Palestinians making it.

The group tour was set up by Cremisan's British distributor, Della Shenton. I introduced myself as people took pictures and listened to Kokaly describe the winery. Shenton, in her fifties, wore big glasses and had a friendly, to-the-point personality. In 2005 she began selling Cremisan wines to restaurants, stores, and churches in the United Kingdom, when the monastery started exporting beyond the Middle East. Shenton acknowledged that in the beginning Cremisan had many problems filling orders: "We'd be promised the shipment, say, in February and it would arrive in June."

The winery tour finished and we moved to the new gift shop, packed with people tasting samples and buying bottles. "I feel far more confident now than I did," Shenton continued. "I've seen the quality improve enormously. I can now confidently spread the word in England. Until now I've tended not to do a proactive marketing. People have come to me." I mentioned Ottolenghi, the famous restaurant whose sommelier had praised Cremisan wines. "Yes, I sell to them," she said with a smile.

I launched into my rambling story of accidentally tasting the wine in the hotel room. After so many years I didn't really expect to understand why the mysterious old red vintage was so distinctive, while the new red isn't. It turned out that Shenton had known Father Ermenegildo Lamon, the old winemaker, who had come to the region from Italy in 1949. He learned about the local grapes from an even older generation.

"He was a wonderful, elderly Salesian, who was the winemaker and the wine master. And what he didn't know about wine growing here and wine production here wasn't worth knowing. However, he kept all his knowledge in his head. No spreadsheets, there was nothing written down," she said with regret in her voice. "But it of course meant that it was difficult to pass this information on. And sadly, when he became ill, he developed Alzheimer's, then Parkinson's, [and] had to go back to Italy. That was a very tough time for the winery. They actually had to throw away *a lot* of wine. Because they didn't know what the heck was in the barrels." Riccardo Cotarella and the Italian winemakers came in with a whole package of support and funding, saving Cremisan. But some traditions died, too. Monks no longer made the wine.

These pieces of information made me almost woozy. I now understood why the red wine from my hotel room didn't taste like Cremisan's new red. Father Lamon never got to pass his knowledge on to Laith Kokaly and Fadi Batarseh, today's winemakers. All the DNA, mass spectrometry, and liquid chromatography testing in the world wouldn't bring back the wine I'd tasted. Father Lamon made wine for more than fifty years, and as far as I can tell, not a single wine critic or publication ever interviewed him.

I realized where the earthy flavors I'd tasted years before had gone, too. Cremisan's old fermentation vats were lined in concrete—a once popular style that went out of fashion when stainless steel arrived. Concrete breathes differently than steel, and can become a home to distinctive microbial communities. It gives wine an almost amphora- or qvevri-like minerality. Some boutique California wineries are now going back to concrete fermentation.

But I learned that Cremisan stopped using the concrete vats after Cotarella began advising, removing that distinctive taste. Shenton told me the old winemaker had also done a lot of blending with grapes such as Alicante, which could explain the spicy notes I remembered. The new Cremisan red didn't use that grape.

My hotel room wine was gone, and I felt . . . relieved. All the reading I'd done on the psychology of taste suggested that remembering flavors is tricky, anyway. Yet finally I knew that the wine had actually changed, and I had, too. I now paid attention to details such as the concrete vats. Putting all the pieces together, I recalled something from the previous spring that hadn't made sense at the time.

Winemaker Yiftah Perets works at Binyamina Winery, one of Israel's largest. He loves ancient grape history, too, and has studied the Cremisan grapes, Israel's wild grapes, and archaeological wine remains. Perets had spoken excitedly of trying to taste wines and grapes that King David, Socrates, or others drank in the past. "Wine is a free spirit and it has its own rules. For me it's a piece of history. And this is what drives me. I found the past very relevant—always. [But] for me it's not the past, it's the future."

That clicked. Without realizing it, my thoughts were working in the same way as they bounced from the six-thousand-year-old Armenian cave to questions about present and future wines. Understanding the past was my way of understanding the present.

I said good-bye to everyone at Cremisan, headed back to Jerusalem, and prepared to visit the ancient city of Gath. I wanted to see how archaeobiology works in the field.

Aren Maeir walked up a modest, dusty, and to my eyes, seemingly barren slope. "We're standing on what probably [are] the remains of one tower of the city gate. That means just below the surface, there are Iron Age remains just waiting to be excavated," Maeir said. Parts of Gath, Goliath's supposed hometown, were under my feet. History, buried beneath millennia of dirt and dust. The only sound was the background buzz of insects, birds, and a far-off power plant. Blue sky silhouetted a hill where the Philistines flourished four thousand years ago.

As we walked around the mostly hidden ruins of Gath's lower city, I asked Maeir how hi-tech tools had changed his work. He likened basic archaeology to a jigsaw puzzle with ten thousand pieces—but you only had three hundred of them. And there was no picture on the cover of the box to go by, he added with a laugh. It's like comparing medicine of the 1800s to the present. In both cases doctors want to help their patients, but now MRIs can pinpoint disease.

"Some of the things we're excavating [now] we didn't even know existed twenty or thirty years ago," Maeir said. "We've moved into being able to look at the micro-view of the ancient remains. I would say our image of the Philistines as the archenemies of the Israelite Judites is perhaps not exactly the case. Right over here, we excavated a temple, from the tenth or ninth century [BC]." Maeir pointed to what looked like a hole in the ground next to an old stone wall. "Right next to this altar, among all the various objects that were offered to this temple, we found a jar which was made of clay from the Jerusalem area. And on this jar there was an inscription, a name, a Judite name. So that means that probably someone from the area of Jerusalem brought a jar to this Philistine temple." The telling detail of the clay's origin came from a process called thin section petrography, which uses a tiny sample slice to analyze its structure.

Teams at Gath have found sets of wine vessels from the city's heyday in about 2000 BC, along with many artifacts that suggest commerce and daily life in the distant past were more complex than we had known. Researchers used mass spectrometry to analyze the Philistine plaster, and were surprised to find that it came from Greece. DNA samples from ancient pig bones yielded a whole new narrative, too. "The pigs the Philistines used were pigs that were [originally] brought from Europe," Maeir said. "We would have had no ability to know that [before]. Things like this really add depth and color—put some flesh on the bones—of the story we're telling."

In 2013 Israeli researchers found traces of cinnamon and other exotic spices on three-thousand-year-old pottery at Dor, a coastal city that traded with Egypt, Greece, and other nations. Maeir said the molecular analysis precisely matched the residue to cinnamon from India, even though it can be found much closer, in Africa. "We have this image of people in antiquity living just in their little village in their immediate surroundings. But things came from really far away—much farther than we assumed. So there's a very, very long-range trade in various types of materials, which played an important role in daily life, and in ritual."

We walked down the slope towards a dry streambed. "I think the parallel perhaps is the Israelis and the Palestinians today. On the one hand, there is a conflict. But on the other hand we live together, we work together, we eat the same food, we dress the same way, we have the same humor," Maeir said.

Mayer-Chissick expressed the same sentiment, and I thought that over after I left Gath and returned home. All of those years earlier, why had I so severely questioned *The Oxford Companion to Wine* claim that the Holy Land lacked vineyards for most of the last fourteen hundred years? Perhaps because I knew from all my travels that people in the Middle East, and all over the world, aren't so one-dimensional.

Much of the Cremisan mystery was solved, at least in my mind, though Shivi Drori's team still hadn't published its final scientific paper on how those grapes might fit into the family tree of wine grape evolution. I would wait for that answer, but now it was time for the next stop on wine's migratory route: Italy.

11

Italy, Leonardo, and Natural Wine

I've drunk Sicilia's crimson wine!
The blazing vintage pressed
From grapes on Etna's breast,
What time the mellowing autumn sun did shine:
I've drunk the wine!

—BAYARD TAYLOR, "SICILIAN WINE," 1854

Winemaking first came to Italy about 3,500 years ago, moving west after being established in Greece. Sicily was probably one of the first regions with major production, so I made an appointment to visit Arianna Occhipinti. In an industry filled with seemingly endless profiles of crusty male winemakers, it's easy to see why many people welcomed a twenty-something woman with a gift for making bright red Sicilian wines with depth, freshness, and a sense of place. Wine magazines sometimes photograph Occhipinti looking like a model, her long, wavy, black hair cascading around a gaze that varies from intense to seductive. For one shoot she perched on a wine barrel with an old limestone wall in the background; in another she bent joyously down in the vineyard dirt, smiling at the camera with youthful exuberance.

I first tried her wines at a Raleigh, North Carolina, tasting— evidence that her fame has spread beyond the major cities on both US coasts. Critics around the world have praised Occhipinti's wines, which are made with traditional Sicilian grapes such as Nero d'Avola. Her success is part of a trend, too. Italy has done more to identify, protect, and promote local grapes than any other country in the world. Over four hundred varieties are now used to make wine (compared to about ninety in Spain) and regional chefs and food critics promote once obscure varieties such as Cococciola, Dindarella, Pecorino, and Vespolina. The *Wine Grapes* authors noted that "there are initiatives all over the country to save from extinction and recuperate historic vine varieties that until recently have been known to perhaps only one or two farmers."

A visit to southeastern Sicily told another part of the story. Driving around country road SP-68 looking for Occhipinti's vine- yard, I saw some healthy vineyards but also barbed wire fences guarding fields of dead vines. There were more than a few old, ruined, stone farmhouses. A stray dog picked through a pile of trash. The region was struggling, not booming. It was a testament to how much Occhipinti had accomplished, more telling than the glossy wine-centric photo ops.

I found her winery behind a stone entrance with remote- controlled iron gates linked to an intercom. Men on heavy equip- ment were working on a new white-walled storage structure. A perfectly renovated farmhouse with limestone walls and a terra- cotta roof stood next to a new winery building.

Occhipinti was sixteen when her uncle Giusto took her to the

2000 Vinitaly wine expo. Giusto was a pioneer, too, co-founding COS winery in the 1980s, which emphasizes local grapes and ferments them in clay amphora. It's one of the largest attempts at re-creating ancient winemaking styles in the world. Occhipinti loved the people at Vinitaly and the nonstop wine discussions, and she soon joined the University of Milan's winemaking program. But many of the courses there stressed laboratory work and industrial wine production, and Occhipinti rebelled. In 2003 she sent a manifesto letter to Luigi "Gino" Veronelli, a pioneering wine journalist and advocate for local agriculture. She wrote to him about listening to Nick Cave and the Bad Seeds, reading the French biodynamic winemaker Nicolas Joly, and of feeling alienated from the coursework:

> I'm studying Oenology and Viticulture in Milan. Every day, I experience false winemaking, weighed down by the obvious pressure of industrial forces. I don't like seeing my colleagues learning the wrong methods because they are the winemakers of the future. I don't know if there will be enough time to change their minds, for them to understand that wine is not "constructed" by impersonal or distant hands. Wine must be accompanied . . . I write these words as a young discoverer, to remember that there is still room for a different path.

Veronelli encouraged her vision, and Occhipinti started making wine. At the age of twenty-five she took out a 150,000 euro loan to buy land, and hired other young idealists to work with her. Critics loved her wines, and the business grew and grew. In 2013, she built a new winery with a system to cool the giant storage tanks on hot summer days. She also sells olive oil produced on the farm.

I had an afternoon appointment but decided to stop by in the morning to take some pictures. Perhaps that was unintentionally rude, but by chance I saw what a day at the winery is really like. A small tractor pulled up, towing a cart stacked high with about seventy-five red plastic cartons filled with purple grapes. Occhipinti and three young workers rapidly unloaded them onto a conveyor belt. She darted from side to side, tasting samples, discarding occasional bunches with rot, and adjusting the belt speed. It was hot, water and dirt stained her cargo pants, and her hair was swept up in a working bun. The grapes tumbled into a mechanical de-stemmer and whooshed down large tubes into the cool winery vats. The crew unloaded all the grapes in about half an hour, and finished up by running each crate through an outdoor rinsing system set up next to an olive tree bursting with small green fruit. Occhipinti rushed off to attend to more business, and I thought of an interview she'd given a few years earlier:

> Last August it was my thirtieth birthday, and waking up still wearing my party dress (a very Sicilian Dolce & Gabbana red silk sheath dress with citrus fruits . . .) I felt different, as I finally discovered myself through the wine. The time of the rebel girl fighting against conventions was gone, and I felt more like a young woman, with a precise idea of my wines and my future, maybe with a more feminine touch even if I always say, "I am a farmer, and proud of it." For me going to the vineyards in the early morning is an amazing sensation of freedom, and I realized that my "allure" as a vintner has

nothing to do with the delicate female attitude of gardening, but is more about repairing the tractor or fixing the bottling machine.

I left the winery to have lunch in a nearby tiny, sleepy village with one-way streets, then headed back to talk with Occhipinti. Inside the old farmhouse the high-beamed office space had a Silicon Valley feel—crisp, open, with a long communal table. Drawings and ideas crowded the chalkboards along one wall. One showed an old, bent olive tree growing out of a circle of stones, with two little field mice hiding near the ground. Next to the tree was a bottle labeled *"VINO BUONO"* and next to that an arrow pointing to the bottle topped by the words *"L'olio puro! BUONO!!!"* A small blue plastic coffeemaker sat on a wooden table underneath the board, next to an award from Düsseldorf's ProWein trade fair event, which read "2013 Arianna Occhipinti Newcomer of the Year."

Occhipinti strode into the office, friendly but intense, dressed in a black T-shirt and tan cargo pants, and sprawled in a chair. I mentioned the abandoned vineyards nearby and then asked, "Did people encourage you in the beginning? Did they say 'You're crazy, it's going down?'" She paused for a second. The ruined wineries were ghosts of a boom dating to the early 1900s, she said. "The situation now is many people are [still] abandoning vineyards. And when I started making wine, maybe it was a little bit worse, in the sense that now some wineries are replanting vineyards, something is moving again."

The old wineries were called *palmentos*, and they used a system of gravity and containers carved out of limestone to crush grapes and ferment juice. Another part of Occhipinti's restored farmhouse has an example. "Many of these kinds of farms, the old palmento, were built because they needed to make a lot of wine." The surge in demand came after phylloxera destroyed many older French and European vineyards. Now there are ruined farmhouses all over the region. Most vineyard production stopped with the Second World War, and many people turned to olives or greenhouse fruits and vegetables. "So many things are different now," she reflected, gesturing with her hands.

"People sometimes think organic means everything is simple and natural," I said. "But you spent a lot of money on the cooling system. You are really paying attention to details and quality control. Did you want to mix traditional and modern winemaking?"

"This is a good question," she said, sitting up in the chair. "Because my life in wine of course did not start with the cellar that you see now. I started in a small place, in another winery. It was not mine, of course, because I had nothing. So I rented. But it was fantastic because I studied [everything] in a very good way. And step by step I became crazy about logistics."

She spent years making wine with basic, minimal equipment, and the current winery emerged out of all that she learned. It is designed to control production at every point after harvest. "It's important because two days ago was like 35 degrees Celsius [about 95 degrees Fahrenheit] outside. We are not in the Loire. We are lower than Tunisia in latitude," she said of southern Sicily. Just a few days of excessive temperatures can lead to wines with burnt flavors.

"You were saying people think organic farming is simple. This kind of farming is not easy. It is really difficult. There were many sacrifices. [But] I'm a person who doesn't stop if I find some wall. I just try to maybe climb, maybe go around, but I want to go over.

Always. Always." She paused. "Now everything is working really well. People are happy. This is my life, my farm."

Occhipinti isn't alone in seeking a business that brings traditional farming and winemaking methods into the twenty-first century. Since 2010 Italy has launched more than a dozen national sustainability programs for winemakers, offering agricultural, academic, and marketing research to regional businesses. Various programs encourage reducing the use of pesticides and fertilizers, making wine without adding chemicals, supporting crop biodiversity, and water conservation.

Now Occhipinti is experimenting with other crops. She planted a local strain of wheat, to grind into flour. She bought some land in mountainous areas, to make a different type of white wine. And she wants to restore more of the nearby abandoned vineyards. "So it's farm dreams. My dreams are in agriculture. Not other things," she said.

I asked Occhipinti if she thinks other countries will grow beyond focusing on wine made from just a half-dozen famous grapes. "It's going to change, I'm sure," she said, but acknowledged that changing palates is a slow process.

Some critics have turned on so-called natural wine, because they feel too many bottles exhibit sour or spoiled tastes. There's no question that such wines are sensitive, and perhaps harder to produce. But Occhipinti has succeeded in producing consistently fine wine, and so has Alaverdi. I wondered what scientists had to say about the fierce debate over natural wine, which is complicated by profound questions over exactly what the term means.

While there's no solid evidence that heavy pesticide or chemical use leads to significant concentrations of harmful chemicals in your wine, they do kill off beneficial microbes, insects, and nearby plants, and may pollute nearby streams with runoff. That alone is worth avoiding. "Wine and table grapes currently receive intense chemical applications to combat severe pathogen pressures," Sean Myles noted in one of his papers.

Isabelle Legeron is often seen as the leader of the natural movement. The founder of the RAW WINE fair, now held in various countries including the United States, she is a French winemaker and Master of Wine who comes from a family that has produced Cognac for six generations. In a 2011 essay first published in *Decanter*, she wrote that natural wine isn't a fad. "Natural wines . . . have existed since time immemorial. When wine was first made 8,000 years ago, it was not made using packets of yeasts, vitamins, enzymes, Mega Purple, reverse osmosis, cryoextraction or powdered tannins—some of the many additives and processes used in winemaking worldwide."

"Natural growers make a panoply of wines, but all share a similar outlook: nurturing biodiversity while embracing and observing nature, rather than fighting to control it," Legeron added. In other words, they avoid or strictly limit the use of pesticides, commercial yeasts, and other additives. Some natural winemakers use clay qvevri or amphora, but many don't.

Cecilia Díaz, a German environmental and food scientist, analyzed twenty different wines from various European countries made in clay vessels, and compared the levels of organic acids and minerals such as calcium and phosphorus to wines made using modern methods. "[The clay wines] contained higher levels of antioxidants and total phenolics, the latter up to 10 times more abundant in the white wine varieties than in conventional wines."

Despite being fermented in mineral-rich clay, the mineral content fell within normal ranges, and phosphorus levels were only slightly higher.

In related research Díaz found that white wines made using traditional methods (skin contact, clay fermentation) "have higher antioxidant properties than commercial white wines . . . Total antioxidant status (TAS), showed average values four times higher in wines produced in amphoras than commercial white wines." Only white wines produced in traditional clay showed tannin levels similar to red wines. Those wines also had more resveratrol, a compound that may have health benefits.

Roberto Ferrarini, an Italian scientist, found differences in Georgian wines, too. Prolonged contact with grape skins led to wine "enriched with phenolic compounds, such as catechins, proanthocyanidins, [and] cinnamic acids . . . Additionally, the aromatic composition is completely different from that of conventionally produced wines."

It's clear that natural or traditional winemaking has intrinsic differences. Some critics claim that natural wines are unstable and prone to spoilage or off-flavors. The critic Simon Woolf took a deeper look at Georgian natural wine. "You might think that this laissez-faire production method would be prone to flaws—how hygienic is a clay amphora, and what about oxidation? Well, that's the interesting part." Woolf reviewed his tasting notes for more than fifty qvevri wines, and was surprised to find that the well-made ones do not have a pronounced oxidized taste (an overall dullness of flavor that comes when wine is exposed to air for too long). "Yes, the aromas can be surprising and dense—cooked fruits, honey, jasmine, herbs and floral notes are all common in the whites—but they are quite distinct" from intentionally oxidized wine like sherry, Woolf said, adding that many traditional Georgian red wines use no added sulfur dioxide at all. (That's an antimicrobial preservative added to most wines around the world, in varying doses.)

I asked Jamie Goode, the British wine critic and scientist, how he views the topic. "You need to be very skilled to make good natural wine. You need much more skill working without the safety net of sulfites and other things than you would if you were just making a conventional wine," he told me. "The really good people are on the ball. They know what to do. They know about microbes. I think that's the key for the ones who make good natural wines . . . they're working meticulously."

Goode also had an insight about what technology has and hasn't done for winemaking. "You don't need any technology to make great wines. You can make great wine very simply, with techniques that would have been available a hundred years ago. I think that's a good point. There's no reason why any of the wine today would actually be any better today at the high end because of technology. But technology would certainly help make good wines more consistent."

The influential American critic Robert Parker isn't opposed to natural wine principles, he just scoffs at the word. In 2014 Parker addressed a California wine conference and mentioned some of his long-held rules for exceptional wine: that it allows expression of the vineyard terroir, micro-climate, and purity of the grape variety; that the wine isn't excessively manipulated; and that it "follow[s] an uncompromising, non-interventionistic winemaking philosophy that eschews the food-processing industrial mind-set of high-tech winemaking. In short, give the wine a chance to make itself naturally without—ah, I hate the word naturally!—without the human element attempting to sculpture or alter the wine's intrinsic character."

I couldn't agree more. A "natural" claim is no guarantee of quality, but we need more winemakers who pay obsessive attention to their land, vines, and wines, not more pesticides or machines. I want people and vineyards such as Occhipinti, Alaverdi, and Cremisan to succeed, not be buried by the clever marketing of international wine corporations. To me, that's the bigger battle: not just natural wine versus the world, but small, dedicated, regional winemakers versus the corporations.

SICILY IS A winemaking world unto itself, so I wanted to visit someone on the mainland of Italy. Elisabetta Foradori was on my list for her pioneering work to re-introduce clay amphora wine, with a modern touch. Her winery is near the Swiss border, but she had a last-minute scheduling glitch and our appointment fell through. That left me in Milan with a couple of open days. I scanned local papers and chanced on an unusual opportunity: using archaeological and historical evidence, scientists had re-created a vineyard Leonardo da Vinci once owned, in land right across the street from the Santa Maria delle Grazie church, where he painted *The Last Supper* mural. The vineyard land was reportedly part payment for the painting.

Serena Imazio, a grape scientist who worked on the vineyard project, met me on a busy street that was once on the outskirts of Milan. We walked through a building to a long, narrow garden that stretched back for a few hundred feet. I could see the spire of the Santa Maria church across the street, just as Leonardo would have.

"It was a crazy idea in the beginning. I thought, OK, they are all crazy," Imazio told me. "Just because Leonardo da Vinci was here in the sixteenth century, it's impossible to trace back all that story, even using genetics, to find something related to that vineyard." She's a biologist who specializes in grape genetics, particularly how grapes became domesticated as the knowledge of winemaking moved across the Mediterranean, from east to west. Imazio took me back to the beginning of the project, long before "Leonardo's Vineyard" was opened to the public in 2015.

Leonardo moved to Milan in the 1480s. In 1495 Ludovico Maria Sforza, the Duke of Milan, commissioned *The Last Supper*. Leonardo finished the painting in 1497, and the Duke gave him a sixteen-perch vineyard that year. "Perch" is an old European land-measuring term that dates back to Roman times. It is a little hard to pinpoint the size because the various kingdoms measured differently. One scholar estimated the vineyard was two or three acres, about 165 by 660 feet. Leonardo kept detailed notes about the value of the vineyard in his notebooks, suggesting that it was worth just over 1,931 gold ducats. At the time a horse cost about forty ducats, and a government employee was paid about three hundred ducats per year.

A sixteenth-century map of Milan shows the vineyard with one tree in the center of the walled garden and additional fields nearby. Leonardo's property included a small house, either a residence or perhaps a farm building. Imazio said the vineyard gift was important to Leonardo because only landowners could be citizens of Milan at the time. The French invaded the city in the summer of 1499, deposed the Duke, and ultimately confiscated the vineyard, so Leonardo left for Venice. In 1507 the French asked him to return to Milan and formally returned the vineyard property.

Leonardo appears to have lived in Milan for a few years before leaving permanently in 1513. He died in France in 1519. His will left equal shares of the vineyard to his apprentice and alleged lover Gian Giacomo Caprotti, known as Salaì, and his servant Giovanbattista Villani. Over the centuries the house fell into ruin. In 1920 a new owner began restorations, and took pictures of the vineyards. A fire, urban sprawl, and Allied bombings during World War II brought new devastation. The vineyard disappeared.

Imazio came to her part in the story as we walked along a gravel path. In 2014 a local historian and wine lover noticed a small plaque outside the home. It said that Leonardo once had a vineyard there. He knocked on the door. The owners confirmed the details, and the historian said, "Really? Nobody knows this story. We have to tell this story." A project was launched to look for archaeological evidence of the old vineyard, more than seventy years after the last vines had died. Imazio worked with Attilio Scienza of the University of Milan, a renowned scientist who's published more than two hundred papers on grapes and winemaking. He's a leading supporter of native grape varieties, and was one of Arianna Occhipinti's teachers. According to Imazio, Scienza tells everyone, "This is our competitive advantage against the popular international grapes, such as Chardonnay, Merlot, and Cabernet Sauvignon."

"You were skeptical about finding evidence here?" I said, responding to Imazio's expression.

"Very skeptical," she said. "I'm a biologist. Mainly I thought we would never be able to find exactly the places where the roots of these plants were. [But] the people living in this house over the centuries were aware that Leonardo's vineyard was there. And incredibly this part of Milan wasn't built until the 1920s. We have lots of pictures showing where exactly the plants were. So we started this work, going down and excavating. And we found some little, very small traces of woods and seeds, that could be useful for our story."

The scientists continued excavating, finding all sorts of debris and plant remains. These could have come from any species, so they spent months carefully cleaning the material. Imazio explained what happened after that. "First, we tried to understand if we could find some DNA inside. And we found something, but we didn't know what DNA it was. I thought, OK, we found DNA, but it's not grapevine." A process called whole genome amplification was the next step. "That means you can repair some pieces of the DNA you find," she said. The results showed remains from the *Vitis* genus—grapes. Imazio thought "OK, [but] the *Vitis* genus is full of species." Another round of genome amplification uncovered a few samples of *Vitis vinifera*, the domesticated grape, and a DNA database search showed a probable match with Malvasia di Candia Aromatica, a widely planted Italian white grape with a long history. The organizers planted seedlings of Malvasia in 2015.

It was an impressive outcome, and I evaluated the details as we walked through the gardens next to the tiny vineyard. A group of tourists chattered nearby and the noise of cars and construction wafted in from surrounding streets. Imazio reminded me that Leonardo grew up in the country village of Vinci, which was full of olive groves and vineyards. She thinks he rested in the Milan vineyard after working on *The Last Supper*. There are frequent references to "vermillion wine" in Leonardo's notes and receipts. One country fable he wrote in the 1490s mentions "a spider's web in a vineyard," and *The Last Supper* shows small glasses of red wine on the table in front of each disciple.

In the end, Imazio was surprised by the findings. "It was also a sort of lesson for me about skepticism." She had underestimated how much could be learned from tiny fragments of plant material, though such potential is part of what attracted her to grape genetics in the first place. "[In DNA] I found this beautiful story about going back to the origin of grapes," she said. That information could help trace and understand human migrations. "When you study the genetics of grapevines, one of the most interesting things is to see how, in the Mediterranean, you have exactly the same variety of grapes in different harbors. So you have Malvasia, the same variety, the same genotype in Croatia, you have it in Greece, you have it in the south islands of Italy, you have it in Sardinia, and then you have it in Spain." Imazio said the DNA evidence illuminates and supports the theory that winemaking spread across the Mediterranean by showing the routes that Phoenician ships took as they explored, spreading vines and knowledge of winemaking along the way. "It really gives you an idea about the lives of human beings thousands of years ago."

Imazio has also worked with Georgian researchers to understand how domesticated grapes moved out of the Caucasus and across the Mediterranean. "What is nice is to see how the germ plasm belonging to the Georgian *sylvestris* somehow arrived in Europe and gave a sort of print to vineyards across the continent," she told me. "It has in some way contributed to building the genetic material that we use in our cultivation nowadays."

Imazio knows and respects José Vouillamoz, but thinks it will be difficult to find a single domestication point in the Caucasus from eight or ten thousand years ago. The pieces of Caucasus grape DNA grow fainter and fainter in vine genomes the farther west researchers look. Perhaps several varieties were domesticated around the same time, in different regions. There could be more than one Holy Grail of wine grapes. Thousands of years of wild and human-induced crossbreeding makes it hard to unravel the truth, especially given the scarcity of funding.

I felt Imazio and her colleagues had uncovered convincing evidence of Leonardo's vineyard, but found no conclusive proof his grape was Malvasia di Candia Aromatica. Perhaps it was planted long after Leonardo died. It's still a great project.

That night in Milan I did some more reading about Leonardo. *The Last Supper* is one of the most famous paintings in the world, a scene that captures love, fear, betrayal, and forgiveness in the various faces at the table with otherworldly skill. There's just one thing: Whether you believe Jesus was a real person or a character in profound stories from two thousand years ago, those people didn't sit at tables. They sat on cushions, Middle Eastern–style.

France was my next stop.

TASTINGS

You could spend a lifetime exploring all the different local wine grape varieties in Italy. One way to start is by picking a region, for example comparing Sicilian wines to those from the north.

Occhipinti, Vittoria, Sicily

SP68 (red, from a blend of Frappato and Nero d'Avola grapes)

SP68 (white, from a blend of Albanello and Zibibbo grapes)

COS, Vittoria, Sicily

Zibibbo in Pithos (white; amphora-aged)

Pithos Rosso (red, from a blend of Nero d'Avola and Frappato grapes; amphora-aged)

Foradori, Mezzolombardo, Italy

Fontanasanta Nosiola (white, from Nosiola grapes; amphora-aged)

Granato (red, from Teroldego grapes; barrel-aged)

The story of how Malvasia white grapes spread across the Mediterranean intrigued me, so here are suggestions for a connect-the-dots Malvasia tasting. The producers go from east to west, just as winemaking did several thousand years ago—with one in California to cap it off.

Domaine Douloufakis, Crete, Greece

Femina (white)

Kozlović Winery, Croatia

Malvazija (white)

Monastero Suore Cistercensi (Monastic Order of Cistercian Nuns), Lazio, Italy

Coenobium Ruscum ("orange," from a blend of Malvasia and two other grapes)

Blandy's Madeira, Portugal

Vintage Malmsey (white; if you want to splurge, an old bottle of classic Madeira can cost thousands of dollars)

Los Bermejos, Canary Islands, Spain

Malvasía Naturalmente Dulce NV ("orange")

Birichino, Santa Cruz, California

Malvasia Bianca (white)

12

Wine and Foie Gras

When I praised the wine and asked him what it was,
he said simply, "C'est du vin de ma mère [my mother's wine]!"
—HENRY JAMES, *A LITTLE TOUR IN FRANCE*, 1884

A huge picture of two smiling, middle-aged women greeted me at the Bordeaux airport. They stood in a vineyard holding wine bottles, next to a slogan proclaiming:

SO ORGANIC

SO HAPPY

I'd expected wine PR, but not that message. The women, surrounded by leafy greenery and blue skies, were dressed for fieldwork. The vineyard's name was in English, French, and Chinese. Someone was attempting a Bordeaux makeover, which made me question how many people in a city known as the Wall Street of the wine business really supported organic farming. Perhaps that's why I felt subversive. Bordeaux wasn't my primary destination, nor were Burgundy, Loire, or Champagne. No, I was in France to see the lesser-known regions, almost like a French tourist visiting New York City to see Staten Island and Queens. But not quite.

Archaeological and DNA evidence show that French winemaking started along the Mediterranean coast, from Marseille to Spain, not in the now-famous inland wine meccas. Until about 1000 BC the Gauls managed with beer and mead-like beverages. But the seafaring Phoenicians, who took winemaking from the coasts of Lebanon to Greece, probably spread the knowledge to Italy next, then France and North Africa. By 500 BC the Gauls had realized making wine themselves was more profitable than importing boatloads of amphoras.

In my travels I'd seen winemakers who were embracing local grapes and shunning famous ones, but what about France itself? Just as in other countries, many old, native grape varieties were dug up and discarded in favor of popular ones. But it turned out some French winemakers suffered from Chardonnay and Merlot fatigue, too.

A whole region in southwest France has successfully returned to making wine with their native grapes. I was headed to the Plaimont wine cooperative in Gascony, near the border with Spain—the French side of the Pyrenees Mountains—to hear their story. Plaimont markets the grapes of the region on a far larger scale than any of the wineries I'd visited so far. Could an heirloom wine for the masses exist without squelching distinctive tastes? I was a little apprehensive. What if Plaimont didn't measure up? What if the classic grapes really are superior?

I left Bordeaux early the next morning, thinking the GPS directions looked easy. After driving south for about an hour and a half on a highway with almost no traffic, the Pyrenees rose up on the horizon like dark clouds. I took the suggested exit, and soon the zig-zaggy roads were impossibly narrow, with room for just one car. Was I lost? The rolling land alternated between corn, livestock, vineyards, and little valleys of forest. Most of the neat one-story stone farmhouses appeared to be lived-in homes, unlike some touristy areas down on the coast around Marseille.

Suddenly I arrived at the tiny village of Saint-Mont. It was 8:40 a.m., seemingly the perfect time to get coffee before my meetings. Wrong. There is no coffee shop in Saint-Mont, just a municipal building and a children's school. So instead I took a deep breath of the fresh country air, filled with a myriad of scents, and contemplated the former Benedictine monastery, dating to 1050, that dominated a nearby hill. A big farm combine rumbled by, and I saw how different Gascony is from the more populous and urbane Bordeaux.

I met my tour guide Diane Caillard at Plaimont, which is just a minute from the town center. She grew up in southeastern France but fell in love with this area. "It's very hidden and people are really fighting to stay together, to stay here and make the region more successful," she said.

Plaimont was founded in the early 1970s by local winemaker André Dubosc, who sensed trouble. Gascony was famous for producing Armagnac, France's oldest named brandy. A 1310 manuscript by the Franciscan Vital du Four claimed Armagnac had forty medical and spiritual virtues, including stopping tears, helping memory, inspiring joy, and promoting witty conversation. But after World War II Armagnac had trouble competing with the better-known Cognac. Many vineyards were ruined or neglected during the war, along with the caves traditionally used for aging brandy. Money was tough to come by in the post-war years and farmers made deals with big producers to rip up the Armagnac grapes and replace them with high-yield varieties.

Some consultants told Gascony farmers to join the Merlot–Cabernet stampede, but Dubosc worried about that, too. How could they compete with Bordeaux and Burgundy? Dubosc urged local vineyards to replant the old varieties. At first many resisted, because they were paid by the kilo. Dubosc told them Plaimont would pay for quality, not quantity. He worked to create new markets, and Plaimont now represents two hundred vineyards and their families. It has grown into a multimillion-dollar business. Dubosc retired in 2006, and the respected wine journalist Tim Atkin later said Dubosc should have run the entire French wine industry, given his visionary work.

My first stop with Caillard was a 190-year-old vineyard that's part of Plaimont. It received French *monument historique* status in 2012—an honor usually given only to buildings. A balding, middle-aged man in coveralls emerged from a barn, wiped his hands, and smiled a shy greeting.

Eight generations ago Jean-Pascal Pédebernade's ancestors planted this small vineyard, leaving room for oxen to move between the rows. In France, vineyards are taxed differently than other land, and old local records show that the first recorded harvest took place here in 1827. The vines themselves could even date to around 1810, when Napoleon ruled Europe. Jean-Pascal's eighty-nine-year-old father still works on the farm, and he tells a story of how his great-great-grandmother listened to one of her ancestors describe seeing the vineyard in the mid-1800s.

We walked up a path. The vineyard looked like a grove of tall bonsai. The vines were about five feet high, but with huge, twisted trunks propped up by pieces of wood, like old people using canes. The leaves were a vibrant, healthy green, and the bunches of grapes looked glorious in the September sun. Yet for decades neighbors thought the old, marginally productive vineyard was odd, if they thought of it at all. For a long time the wine industry either didn't know or didn't care.

In the 1970s and '80s agricultural experts suggested that the Pédebernade family tear up the old vines in exchange for a one-time cash payment—the grubbing up schemes I had learned about from José Vouillamoz. The European Union has spent billions of dollars on similar programs, as part of an effort to stem an embarrassing glut of wine: Each year millions of gallons end up as industrial alcohol, in a highly subsidized and vastly expensive price support program. In 2007 the EU estimated that it spent around half a billion euros every year just getting rid of surplus wine. Between 1988 and 1993 growers, mainly in southern France and southern Italy, were paid to destroy almost eight hundred thousand acres of vineyards—an area roughly equivalent to all the vineyards in America.

The Pédebernade family refused the grubbing-up money. The little vineyard survived.

When scientists analyzed DNA from the six hundred Pédebernade vines, they found more than twenty varieties—seven previously unknown. Those are called Pédebernade Nos. 1 to 7, and the vineyard also includes samples of Claverie, Miousat, Fer, and Canari grapes. Historians are heralding the vineyard as a remarkable example of biodiversity, genetic heritage, and ancestral cultivation methods.

The little plot survived the phylloxera plague of the mid-to-late nineteenth century because of the unusual, sandy soil. Phylloxera are aphid-like creatures that build tiny underground tunnels as part of their reproductive cycle, but those collapsed in the sand, creating a small island of immunity. Old vineyards have faced other challenges, too.

Caillard then took me to meet Plaimont's current director over at another vineyard of Tannat grapes that dates back to 1871. Olivier Bourdet-Pees was casually yet fashionably dressed in a beret. He's still worried about overall vineyard trends in France. In 1950 about 53 percent of all vineyard acreage used just twenty grape varieties. Today, 92 percent is planted with those twenty varieties. "Everyone is producing the same wine with the same tools. I believe that if we get bored with wine, we will stop drinking wine," he said of the threat posed by too much conformity.

Bourdet-Pees said Dubosc did "an amazing job" of persuading local growers to trust the local grapes. "To say to the winegrower, please don't plant this Chardonnay. We will die with that kind of grape. Please, you have to be a little bit patient." No other project of this size makes wine from regional grapes in France, Bourdet-Pees informed me.

Bourdet-Pees stopped near a fig tree at the edge of the vineyard and became animated. He said not so long ago local people thought the Plaimont owners were "Silly men. Madmen. Old-fashioned men. Because it was time to produce a lot with new vineyards. New plantations, new projects, more productive." Bourdet-Pees mimicked some of the comments French agricultural experts made in the 1970s. *Look at this plant. It's completely broken, nearly dead. Why do you want to keep this? If you want to keep this, that means you have a problem. You're not a modern-style winegrower. You have to think another way. This was the vineyard of your grandparent. You have to change it.*

Bourdet-Pees said the family kept the 1871 vineyard for sentimental reasons, because their grandfather and prior generations had made wine from it. For about twenty years Plaimont gave the owners a little bit of money to preserve it, but at first it was more of a museum. "When we started to understand the quality of the terroir, the quality of the soil . . . it [became] quite interesting. We started to think about producing wine," Bourdet-Pees said. It took about four years to bring the vines back to healthy production, and the first vintage was 2011.

"What did you think when you first tried the wine?" I asked.

"The freshness, the roundness of the wine is amazing. I'm not sure if it's because it's ungrafted or because the vineyard is really old. The roots are five meters under the soil. I do think it's really different. We'll try it later and you can see," Bourdet-Pees replied.

I couldn't wait to sample the wines, though the vineyard was glorious. Moss covered parts of the old vines and wasps diligently crawled over ripe figs. I could see the Saint-Mont monastery in the distance. A field of cornstalks was drying just past the vines. The tile roof on a nearby farmhouse was covered in bursts of ochre lichen, its walls framed by plots of flowers and herbs.

Bourdet-Pees reflected on the growing enthusiasm for locally oriented, sustainable agriculture. "It was difficult to believe in this twenty years ago. It was really difficult," he said of Plaimont's idea. "But now many people do want to rediscover old varieties of apples, or tomatoes. I'm pretty sure it is the right moment for us to be discovered again."

There are still challenges. Some agricultural universities in

France use technology to make popular grape varieties fit the land, even when they aren't suitable. Bourdet-Pees said they tell winemakers it is possible to produce wine everywhere, even in the worst conditions, even with the wrong grape, even with the wrong terroir.

The push to plant such grapes is happening everywhere. "If I want to drink a wine from Romania, I don't want to drink Merlot. It's mad," Bourdet-Pees said. "The climate [there] is not so good for Merlot." Over the years Plaimont has found many customers who order well-known wines simply because they know what to expect. They become almost afraid to discover something they have never heard about. I asked what happens when they actually try the Plaimont wines. "They say, can I have another bottle for my friend?"

After a short drive we arrived at Plaimont's beautiful tasting room. It's on the second floor, over a large shop featuring all of their wines. We sat at a circular tasting bar and Bourdet-Pees first poured whites, starting with Plaimont's 2014 Côtes de Gascogne, their affordable table wine. It's made with the Colombard grape, which has a bad reputation in some places, so I didn't know what to expect. The wine was smooth, drinkable, and had an almost startling grapefruit aroma.

"That's very nice," I said. "It still has personality." I'd visited many small, beautiful vineyards, but to be able to make millions of bottles of a distinctive wine is a new challenge. "This aroma, this freshness, makes a difference in the market," Bourdet-Pees said. "Many people can discover it. Not so expensive." Plaimont sells about five million bottles a year of Côtes de Gascogne at about five dollars retail, and I could see the appeal. It was an easy to drink table wine but more interesting than bargain Chardonnays.

We tasted a 2014 Domaine de Cassaigne, made with 70 percent Gros Manseng grapes and 30 percent Colombard. It was a full, long wine that begins with pineapple aromas and progresses to a light hint of oak. "Still exotic, but more maturity," Bourdet-Pees said. It's aged in oak for about six months to give more roundness, and the wine could easily pair with main courses. Many large wineries press grapes and quickly discard the remaining pulp and skin. It can make the whole process, and the wine, somewhat more predictable. This wine was different. "With both Colombard and Gros Manseng the aromatic power is really in the skin. With Colombard it's very long skin contact. About twenty-four to thirty hours before pressing. With Gros Manseng it's six to twelve hours of skin contact."

The next white, Plaimont's 2013 Les Vignes Retrouvées Saint-Mont, was a beguiling blend of grapes. "Gros Manseng is the main part, about 75 percent. The Petit Courbu gives the roundness. And then a wonderful grape, which is Arrufiac. It's spicier, a lot of bitterness. If you are thinking about Arrufiac for 100 percent wine, it's awful. Awful, really," Bourdet-Pees said of the appellation. Plaimont uses four or five percent Arrufiac—like a touch of salt while cooking—in Les Vignes Retrouvées Saint-Mont. The wine had a distinctive, stony profile with aromas of white pepper and a touch of smokiness. Bourdet-Pees said the smokiness emerges only from Gros Manseng grapes that are grown closer to the Pyrenees. It was a remarkable wine, and a triumph of terroir.

The first red wine was a 2011 Monastère de Saint-Mont. It is all Tannat, grown on about five hectares of chalky, clayey monastery land. The first sip was almost like rolling a piece of chalk around in my mouth, but soon clay flavors emerged. "It feels like the taste is stages of terroir. The clay is second," I said, puzzled.

"You can feel it. It's strong," Bourdet-Pees said. "It's OK for me when people say, OK that one I don't like at all. We want to be as close as possible with the quality of the soil. It's fresh, powerful— but it's not so easy."

After a few minutes I took another sip and it was a different wine. The chalk and clay tastes had receded and a lovely cherry aftertaste emerged. "I like this very much, but it still tastes young," I said. Bourdet-Pees told me the wine was aged in oak for fifteen months, and had tremendous aging potential. It was a great example of how important breathing is for complex red wines. Right out of the bottle the Monastère de Saint-Mont was tough, but in just fifteen minutes it blossomed with new flavors. I thought back to the chemistry lesson Marcos Zambartas had given me in Cyprus and now understood it in on a gut level. The molecular bonds in the wine had changed almost before my eyes, revealing new tastes.

The next red, a 2012 Le Faîte Rouge ("The Red Fairy"), was a witty twist on typical blends. It contained 85 percent Tannat, 10 percent Pinec, and 5 percent Cabernet Sauvignon—putting the most famous grapes in a minor, supporting role. It still had a touch of chalky aroma, but earthier, and with more fruit. I asked Bourdet-Pees how long of a decant he recommended. "If you want two hours, three hours, even the next day. If you come back the next day it's even better," he said.

The last wine was a first for me. The 2013 Vignes Préphylloxériques was made from the vines at the 1871 vineyard, the one I'd visited with the fig tree. "I've never had a pre-phylloxera wine," I said. Wine lovers endlessly debate whether the old French vines made better wine, and a few small plots throughout Europe still produce enough grapes to make a vintage. Plaimont made just 1,150 bottles of this one, and few ever go outside of France. I felt very lucky.

Bourdet-Pees poured, and told me it was 100 percent Tannat. I took a sip, getting waves of plum aroma right off the bat, a bit of licorice, and finally green, spicy tobacco. "Wow. It goes on and on and on," I said after tasting dried cherry, as if fruits had actually been aged in the bottle.

Bourdet-Pees smiled like a proud parent. "Some of the grapes are not so mature [when we harvest]. So this profile of tobacco, is from there. But very subtle." It was a magnificent wine, and the taste kept evolving, with apricots and a steely edge showing up after about ten minutes.

"Yeah! A lot of iron. A lot! Very poor soil, very, very poor," Bourdet-Pees said, and we both marveled that the vines can produce a remarkably smooth wine that is full of flavor. By chance I'd experienced one of the mysteries of terroir. Before tasting the Vignes Préphylloxériques all I knew about the land was that it had sandy soil, thus protecting the vines from phylloxera insects. Yet I tasted iron in the wine, and Bourdet-Pees confirmed the soil in that particular vineyard is full of iron. Here's the conundrum: Scientists tell us that minerals in soils don't directly impart flavors to wine grapes, despite myths claiming they do. How, then, did the metallic flavor get in the wine?

The very concept of terroir opens up a scientific and psychological quagmire. Consider: while regions all over the world have distinctive soil types, in most cases those minerally flavors never show up in the wines, even in areas that are rich in iron, clay, granite, or whatever. Yet those very flavors do show up in particular vineyards, even though wines made just a mile apart can differ—some show terroir and some don't, even with the same grape. That's why wine lovers fall so hard for wines with that ephemeral trait. The industry also generally knows how to make wines sweeter, drier, fruitier, and more or less alcoholic. Short of chucking rocks or rock dust into the vats (which is actually done), no one knows how to make a grape, a vineyard, or a wine naturally express a mineral terroir.

How, then, did the metallic iron flavor get in the last Plaimont wine I tasted? Bourdet-Pees had not alerted me to that detail beforehand. I'd never read a review of the particular wine and had no way of knowing the soil type. Yet I tasted something iron-y that Bourdet-Pees immediately recognized. If wines everywhere exhibited terroir, or if no wines did, scientists and wine lovers wouldn't fight so much over what it means.

In any case, Bourdet-Pees was right to be proud. "If you put it in a carafe for three to four hours, it's amazing," he said of how the wine opens up. "You can drink and drink and drink and it's still smooth."

After six very distinct wines we headed to lunch with a few more bottles. The restaurant and inn Auberge de la Bidouze was like stepping into the past, too. It's a simple house with woods shutters that offers a big patio surrounded by flowers or a cozy dining room for meals, set among cornfields just a few minutes from Plaimont.

We had pork chops, pâté, and foie gras, charcuterie, and salad with fresh tomatoes and herbs. I usually don't like pork chops much, but these were succulent and flavorful, made from the farm's own breed of pig. Everything was local, unpretentious, and absolutely sumptuous. André Brulé, the chef, is a large, shy and generous man who also waits the table, as if that's a perfectly normal thing for a chef to do.

After lunch Bourdet-Pees told me that he believed things were changing. "I'm really fond of the new generation. I do think that mine, and maybe a little bit older, was wrong [about wine]. The young ones, not all of them, but the really good young ones, want to see wine differently."

On the way back to Bordeaux from Plaimont I was happy, filled with marvelous wine and food and talk. Still, I didn't want to completely ignore the classic wines. That would be foolish. So the next day I booked a seat on one of the many wine tours in the region. The minivan headed off to Château Smith Haut Lafitte. Its vineyards supposedly date to the 1300s, and the stone manor house to the 1700s. In 1990 Daniel and Florence Cathiard, former members of the French Olympic ski team, bought the chateau. They soon added sustainable and high-tech methods to the traditional parts of the business.

We toured the barrel-making room, which starts with slabs of raw oak and ends with magnificent barrels, their insides toasted to precise levels to impart the proper flavor. The main winemaking area was huge, spotless, and computerized. Smith Haut Lafitte uses a program called Oenoview to analyze the perfect harvest time. Data provided by satellite measures plant emissions related to ripeness, providing a digital map of every few square feet of the vineyard.

Tour guide Alix Ounis told us that twenty years ago Bordeaux chateaus—winemaking estates—were not organic or biodynamic at all. Now more and more are, including Smith Haut Lafitte. They farm organically, use oxen instead of tractors in the vineyards to avoid compacting the soil, and capture some of the winery's CO_2 emissions to reduce their global warming footprint. The CO_2 is turned into bicarbonate of soda, or baking soda, which is used in toothpaste and other products. The Cathiards also sell grape seeds to their daughter's company, which uses them in natural skin care products.

Smith Haut Lafitte has a restaurant and a seventy-two-room five-star hotel, but I settled for a basic tasting. Before the Cathiards took over, the chateau wines had a middling reputation. Critics now rate them very highly. I tried one of their best, a 2012 blend of Sauvignon Blanc, Sauvignon Gris, and Sémillon grapes. It was heavenly. Tremendously full, but smooth and fragrant. Then we tried their 2011 Le Petit Haut Lafitte, a second-tier red. It was nice, but obviously not the best vintage.

The first white was an outstanding example of why so many people lust after Bordeaux wines. I contacted Florence Cathiard to learn more after I returned home. "We know in every single row of the vineyard how ripe the grapes are. We then taste the grapes in each plot and mark the vines which will be harvested the following day," she replied in an email. Then an optical scanning machine in the winery looks for imperfect grapes and culls them out. Visitors like the combined focus on sustainability and wine quality, Chathiard wrote, and I wondered if such technology would become the norm throughout France.

I headed back to my (slightly less posh) lodgings, gratified to have tasted both the Smith Haut Lafitte wines and Plaimont's. Curious whether tourists and locals were much interested in rare varieties, I stopped at Le Wine Bar, which had been recommended to me by some friends. It was on the corner of a quiet street, with high ceilings and elegant curved iron window gates. The afternoon was quiet, so co-owner Delphine Cadei had time to talk. Her family is from Bordeaux, but for many years she lived in other cities and didn't particularly long for home. Cadei said that for a long time Bordeaux was dark, grimy, and not a nice place to live. Parking lots covered the wide stone quays, obscuring the river. Bordeaux had been known as a wine town for centuries, but it had a crass business side, too.

That brought to mind Henry James's reaction to Bordeaux in the 1880s. James reported that he couldn't advise readers where to find fine wine, because "I certainly didn't find it at Bordeaux, where I drank a most vulgar fluid." To be fair James visited not long after phylloxera had ravaged French vineyards, so the wine he drank then might have been made elsewhere, and just rebottled in Bordeaux. Such practices have led to a seemingly endless stream of scandal and legislation. For decades in the early twentieth century, the French colony of Algeria was one of the biggest wine exporters in the world. But much of the product was blended with—and sold as—French wine.

A century after Henry James's observation, Kermit Lynch, in his 1988 travelogue *Adventures on the Wine Route*, poked fun at the mind-boggling number of chateaux in Bordeaux. "The landscape is bestrewn and plumed with them. [. . .] But even the name château is a facade, because many châteaux are nothing but dilapidated sheds in which wine is produced."

Cadei said a civic makeover had changed Bordeaux's old city, transforming it into another popular tourist destination, after Paris. The parking lots vanished and the river quays are now full of artists, visitors, and locals. I told Cadei about my wine travels and asked if her customers were interested in unusual grape varieties or regions, or just the classics. She said tourists ask for Bordeaux wines, but many times local people want to taste wines from other places. That sounded a lot like what Bourdet-Pees from Plaimont had noticed. Some of Cadei's regulars try a different wine on each visit, seeking out unknown countries and vintages.

I ordered a half-glass each of three Bordeaux wines and a charcuterie plate—foie gras, country pâté, and luxuriously fatty speck, or cured ham. The food arrived and everything had a creamy, elegant richness. The whole experience, mixed with the narrow cobblestone streets and old buildings, briefly inspired dreams of moving to France.

I learned of another chapter in wine grape history while in Bordeaux. A new generation of worldly, successful Chinese are embracing Western wine. There were several Chinese couples on my Bordeaux tour, and the Western wine industry has taken note: the wine and drinks magazine *Decanter* launched DecanterChina .com in 2012, and Robert Parker's the *Wine Advocate* hired a Shanghai-based critic in 2015. In a sense the passion for Western wine isn't new; there are historical reports of Chinese emperors importing wine via the Silk Road, starting at least two thousand years ago. This modern frenzy has occasionally tragic results. A billionaire and his son died when their helicopter crashed during a tour of their newly purchased Bordeaux estate. As of 2016 there were more than two hundred thousand acres of vineyards in China, planted mostly with—you guessed it—Cabernet Sauvignon, Cabernet Franc, and Chardonnay, even though agricultural experts warn that some of China's regions really aren't suited to the European grapes.

Though ancient Chinese beverages were more of a rice-based brew than wine, archaeologists are finding parallels to Western rituals. Chinese nobility were buried with provisions of wine for the afterlife, too. One story goes something like this:

About three thousand years ago, along the banks of an ochre-colored Chinese river, slaves toiled to finish a tomb. It was loaded with food, honey, jade, weapons, and even a chariot and horses. Elaborate bronze vessels filled with chrysanthemum-infused wine lined one wall, meant for a future feast that never came to be—at least in the way people of the time intended.

Over time dynasties withered and fell. The tomb was forgotten. Insects, worms, and microbes feasted on the underground banquet, with one exception: the metal tops of the jars corroded and formed an airtight seal. The beverage went through some chemical changes and finally stabilized, waiting for the unexpected chapter in its own afterlife.

Recently archaeologists discovered the tomb and found what some had dreamed of but never really expected to find: liquid inside the bronze containers, the vintage of all vintages. Patrick McGovern got to sniff those curious remains, but professional standards and common sense barred him from taking a sip.

I left France and took some time to reflect during the trip home. I'd visited seven countries and the Palestinian Territories. In each place winemakers and wine lovers had grown weary of Chardonnay, Merlot, and the like, and had independently, without knowing one another, started exploring native grapes. Israeli, Palestinian,

Armenian, Georgian, Cypriot, Greek, Italian, Swiss, and French wine lovers all described the same urge to rediscover the grapes and wines of the past. A sociologist could have a field day analyzing how that feeling manifested itself in so many different cultures at the same time. This new movement doesn't mean you have to give up famous wines and their grapes, any more than going out for Vietnamese precludes steak, vegan, or Italian meals. It just shows that there are many more wine flavors—and stories—to explore.

TASTINGS

France is full of distinctive regional winemakers beyond Burgundy and Bordeaux, if you just look a bit. The Plaimont wines are a terrific bargain but hard to find in the United States. I've listed several here, as well as some wines made from lesser-known grapes suggested by Kermit Lynch, the importer and writer. His store in Berkeley, California, or his online site are great places to look for and learn about French and Italian wine producers: www.kermitlynch.com.

Domaine de l'Alliance, south Bordeaux

Dry whites using Sémillon grapes

Clos Canarelli, Corsica

Amphora-aged red or white using grapes such as Genovese, Carcaghjolu, Paga Debiti

Domaine Comte Abbatucci, Corsica

Reds using Sciaccarellu and Carcajolo Nera grapes

Jean-François Ganevat, Jura, Switzerland, near the French border

Many white and red selections using Savagnin, Trousseau, Gamay, and other grapes

Domaine Les Mille Vignes, southwest France, near the Spanish border

Le Pied des Nymphette (white, from Carignan Blanc and other grapes)

DOMAINE HAUVETTE, NEAR MARSEILLE, FRANCE

Reds, whites, and rosé using Cinsault, Roussanne, and Clairette grapes, among others

Clos Sainte Magdeleine, Provence

Cassis Bel-Arme (white, from 65% Marsanne, 15% Clairette, 15% Ugni Blanc, and 5% Bourboulenc grapes)

PLAIMONT, GASCONY

Côtes de Gascogn (white, from Colombelle grapes)

Les Vignes Retrouvées Saint-Mont (white, from blend of Gros Manseng, Arrufiac, and Petit Courbugrapes grapes)

Les Vignes Retrouvées (white, from Gros Manseng and Petit Courbu grapes)

Domaine de Cassaigne (white, from 70% Gros Manseng and 30% Colombard grapes)

Monastère de Saint-Mont (red, from Tannat grapes)

13

The Science of Terroir

*Wine is one of the most civilized things in the world
and one of the most natural things of the world
that has been brought to the greatest perfection.*
—ERNEST HEMINGWAY, *DEATH IN THE AFTERNOON*, 1932

There's an old story about monks in medieval Burgundy who tasted the dirt to predict the best land for vineyards. It's a vivid, apparently insightful detail, but not in the way you might first think.

"The history of *terroir* itself is often misrepresented. It has not been the basis of French wine for centuries, and it wasn't discovered by monks who noticed that the wines from one vineyard were different from those in another vineyard," historian Rod Phillips wrote in his article "The Myths of French Wine History." Yet I hadn't imagined the iron and then chalk taste in the Plaimont wines. At least I didn't think so.

What does science reveal about terroir debates? Concise answers, or a web of new mysteries? Perhaps both. It's all enough to make a wine lover long for the days when we could just say that rocky soils (think granite) impart rocky flavors in wine.

"It really bothers my wife and me," said Sean Myles, the grape geneticist from Nova Scotia. His wife runs a vineyard. "You go to a winery for a tasting, and someone says, 'Oh yeah, it was grown on slate and very rocky soil, and that's what gives it the minerality.' And we're like, 'No! It doesn't!' It really can't. That's totally oversimplifying how things work. It doesn't pick up minerals from the ground."

Alex Maltman, a geologist at the University of Wales, published a paper on the topic. In it, he states, "The notion of being able to taste the vineyard geology in the wine—*a goût de terroir*—is a romantic notion which makes good journalistic copy and is manifestly a powerful marketing tactic, but it is wholly anecdotal and in any literal way is scientifically impossible."

There is a long history of wine lovers using the term "terroir" in a variety of ways. A 1900 story in the *New York Times* observed that a certain wine had "a gout de terroir almost Burgundian." A 1988 *Times* article about a wine conference corrected that oversimplification. Gérard Seguin, a soil expert from the University of Bordeaux, explained that terroir is often translated to mean simply soil. "[But] the word, properly understood, meant far more: an ecosystem that integrates site, soil composition and drainage, climate, vines, rootstock, grape varieties and the human role in farming and wine making." Then a 1993 *New York Times* wine review went back to the old shorthand by recommending a wine with "a classic Burgundian gout de terroir—taste of the soil—and a lovely bouquet."

A 2014 study of vineyards all over California found that the soils were home to distinct regional communities of microbes, bacteria, and fungi, like various neighborhoods in a city. DNA sequencing revealed that the microbes are still present in crushed Chardonnay and Cabernet Sauvignon grapes at various wineries. The researchers theorized that a "microbial terroir" may explain regional variation among wine grapes. That's important for at least two reasons. First, all those tiny soil creatures influence fermentation.

Second, a microbial definition of terroir suggests a scientific rationale for organic farming. The endless debates over the value of such practices take on another dimension when we look at the entire soil community as a whole. Many other researchers have documented what in retrospect seems obvious: pesticides and fungicides kill off most soil organisms, not just the insects or diseases they target. Our concept of "soil" was far too simplistic. It's not just sandy, rocky, rich, or poor, but a living community. Gardeners know this from cruel experience: certain plants flourish in some soils, yet do poorly in others, and getting the right mix makes all the difference in the world. In a sense, the medieval monks who tasted soil were kind of on the right track. Different soils do produce different wine flavors, just not for the reasons we assumed.

The study looked at the patterns of 273 microbial samples, and concluded they are linked to grape variety, regional weather conditions, and geography. This means it may someday be possible to explain terroir in a way that makes everyone happy. But for now, the battle still rages. In early 2016 the University of California Press published *Terroir and Other Myths of Winemaking* by Mark Matthews, a professor of viticulture at UC Davis. One reviewer called it a "myth-busting book," another came away convinced that "the idea of 'terroir' is likely the most abused and often most useless word in the world of wine." The *SOMM Journal* remarked that it was a "meticulously researched volume that every serious sommelier should read . . . if only to disagree."

The current, fourth edition *Oxford Companion to Wine* broadens the term to "the complete natural environment in which a particular wine is produced, including factors such as the soil, topography, and climate." That's a perfect definition, one that respects both tradition and science. In fact, some ancient writers had similar observations two thousand years ago. Columella, a Roman who was born about AD 4, wrote one of the most important early books on agriculture. His advice rings true even today: "An important consideration is the variety and the habit of the vine which you propose to cultivate, in relation to the conditions of the region. For its cultivation is not the same in every climate and in every soil, nor is there only one variety of that plant; and which kind is best of all is not easy to say, since experience teaches that to every region its own variety is more or less suited."

Another passage could even be relevant to the global use of just a few French and European grape varieties. "Foreign cuttings, transplanted from a different locality, are less at home in our soil than are the native varieties, and for that reason, being strangers, so to speak, they dread a change of climate and situation; and also they offer no definite assurance of quality . . ."

14

Coming Home, and Holy Land Wine

. . . look down from heaven, and behold, and visit this vine . . .
—PSALM 80:14 (KJV)

M y last wine trip to the Middle East ended in the late summer of 2015, seven years after the hotel room tasting. I stopped in Manhattan first and headed downtown on a rainy fall day, to a tasting at Astor Wines & Spirits in Greenwich Village. I went up a flight of stairs into a room filled with sommeliers and wine buyers. They staked out places at tables, sniffing, swirling, and mostly spitting out samples from dozens of bottles, as usual—except all the wines were Georgian.

The chilly rain had fogged the high, arched windows of the historic building, and I tried an unusual glass of white wine. Intensely golden and cloudy, like raw cider, it was wildly different, conjuring up visions of a woodsman drinking in a mountain forest, or perhaps the people in that ancient cave. It came from a small family vineyard, and the label featured a simple drawing of six heads—men and women—with the name "Our Wine." The pourer, a Georgian man, started to explain their wine history. I smiled and said, "Oh, I've been to your country." He looked surprised.

How did such a homespun creation made its way to New York? Seeing Lisa Granik, a Master of Wine who organized the tasting, I asked if this event could have happened ten years ago. "Absolutely not," she said, adding that until recently American consumers weren't ready for such diversity, and Georgian producers didn't grasp our market either. I took another sip of the earthy wine and for a second, it felt like a taste of the past.

Granick is a former Fulbright scholar and Georgetown Law graduate, and one of only a few dozen Americans certified as Masters of Wine, the most prestigious title in the field. Worldwide there are just three hundred and fifty-five MWs, all of whom must pass a difficult multi-year program first developed by London's Worshipful Company of Vintners, a guild that received a Royal Charter in 1363. After the tasting, I called Granik to follow up on why she thought New Yorkers were paying attention to once obscure wines. She said people are seeking new varieties, new producers, and new stories, particularly in urban areas; that meant looking at lesser-known regions. There is a generational shift, too. "[People] are interested in exploring wines that are different from the ones they thought their parents drank." She doesn't expect Rkatsiteli to replace Riesling, but now it has a place in the market.

Months after the event I went to Astor's website, wanting more of the rustic "Our Wine." This is how the staff described it: "Amber-colored and not for the uninitiated . . . Quite tannic and dry as a bone, one could only recommend this wine for the adventurous." The notes add that "If this wine were a pop song it might be called

'Take a Walk on the Wild Side,' 'Crazy,' or perhaps 'Livin' la Vida
Loca!' . . . What you don't get is a light, crisp, fun white quaffer;
what you do get is big, firm, earthy, tannic (yes I said tannic), nutty
monster of a wine that probably lives under your bed!" Then the
notice: out of stock. Eight of the ten Georgian wines Astor listed
at that time were sold out. Granik was right—there are more adventurous wine drinkers in the world than we might think. Now
I was one of them.

In the fall of 2015 the fourth edition of *The Oxford Companion to Wine* came out, with a revision to acknowledge that Israel
does in fact have native grape varieties, and that Cremisan was
making wine from them. After I queried the editors, they also decided to correct the mistaken suggestion that winemaking in the
region vanished for much of the period between AD 636 and the
late 1800s. I admit to feeling vindicated, and suspect future editions will mention Drori's research, and even more Israeli grape
varieties.

Around the same time I wrote a travel story about Cremisan
for the Associated Press. It ran in papers and websites all over the
world, with the headline "Palestinian winemakers preserve ancient
traditions." That November the *New York Times* profiled an Israeli
wine inspired by Shivi Drori's research, and that got even more
attention. Ido Lewinsohn, vintner at Recanati Winery at the time,
explained why they decided to use Arabic, Hebrew, and English lettering on the label for the new wine, stating that the grapes "are not
Israeli; they are not Palestinian. They belong to the region—this is
something beautiful." Lewinsohn later told me that Recanati had
American distribution for the wine, and a red wine made from native grapes is in the works, too. In December CNN.com featured

both the Israeli and Palestinian wines under the headline "What would Jesus drink?"

"Things definitely got pretty busy last fall when the press hit, mostly 'where can I buy it' type of emails," Jason Bajalia, Cremisan's American distributor, wrote me in an email. "We are in pretty much every Israeli restaurant in NYC now. That wouldn't mean anything anywhere else, but there are so many in NYC (and high profile ones) that it is a significant development!"

Cremisan was no longer obscure. I chilled out for a bit, and considered how the hotel room wine ultimately set me off like a viticultural Quixote, traveling ancient wine routes, championing obscure grapes and railing against the glut of famous French varieties. I hadn't seen that coming.

It was time to read and reflect. Rod Phillips, who wrote the essay about French wine myths, had a new book out titled *French Wine: A History*. It had plenty of love and respect for great winemaking, but expanded the myth busting beyond terroir. The reassuring claim that many vineyards in that country go back to the fourteenth or fifteenth century?

It appeals to our sense of lineage, stability, and tradition, all powerful and positive associations. [Yet vineyards] in many French regions were devastated by the Black Death from the mid-1300s and the Hundred Years' War to the mid-1400s, and by the frigid winters of 1693 and especially 1709 . . . [In fact] despite more than 2,000 years of developing a wine industry and wine culture, French producers settled on varieties and locations as recently as many New World producers . . . the idea of an uninterrupted narrative, linking

modern French wine to the distant past, is on very shaky
ground.

And the magnificent, noble grapes the world has fallen so hard
for? "[U]ntil recently, French wines were field blends, made from
several varieties picked simultaneously so that the grapes were
variously green, ripe, overripe, and rotten. They were crushed and
vinified together, the must was fermented in open tanks for weeks,
and the wine was stored in dirty barrels," Phillips wrote.

Well OK, I thought. Are there any other cherished wine myths
left to challenge? One soon came to mind.

I had obsessed over the *Oxford Companion* claim that wine
vanished from the Holy Land for roughly a thousand years, but
kind of ignored the second half of the passage: that Baron Edmond
de Rothschild sparked a winemaking revival there in the 1880s.
There is no question Rothschild had a huge impact on winemaking
and agriculture in what was then called Palestine, but I suddenly
realized his influence wasn't all positive. Rothschild was blinded by
the myth of French noble grapes. In the early 1880s one of Rothschild's wine experts praised the winemaking potential of existing
Palestinian grapes, but the project favored classic French varieties
instead. One historian notes that between 1884 and 1886 there was
"a great deal of uprooting of the ancient indigenous vines." Rothschild had essentially ignored grapes that were perhaps historically
linked to the Jewish religion. More than 120 years passed before
Shivi Drori and his team began studying and saving native vines.
To me the Rothschild wine story seems to illustrate a broader
point: Mizrahi Jews in Israel trace their heritage to other Middle
Eastern countries, Ashkenazi Jews came from Europe, and there is

an ongoing dispute about whether Mizrahi heritage (and links to Middle Eastern food and traditions) are overlooked.

I thought over that contradiction while reading, often with a smile, a collection titled *A Miniature Anthology of Medieval Hebrew Wine Songs*. One of my favorites was by Samuel the Nagid, whose motto seemed to be pray hard but party hard, too:

> Don't speculate on hidden things; leave that
> To God, the Hidden One, whose eyes sees all.
> But send the lass who plays the lute
> To fill the cup with coral drink,
> Put up in kegs in Adam's time,
> Or else just after Noah's flood,
> A pungent wine, like frankincense,
> A glittering wine, like gold and gems,
> Such wine as concubines and queens
> Would bring King David long ago.

Reading the medieval Jewish poets brought me to a final reckoning about Holy Land wine during that era. The biggest exodus of Jews between AD 1000 and 1600 wasn't from Muslim lands, but from Spain and other European countries. Many settled in the Ottoman Empire. In 1453 Sultan Mehmet II openly courted Jewish immigrants with this proclamation: "Who among you of all my people that is with me, may his G-d be with him, let him ascend to Istanbul, the site of my imperial throne. Let him dwell in the best of the land, each beneath his vine and beneath his fig tree, with silver and with gold, with wealth and with cattle. Let him dwell in the land, trade in it, and take possession of it."

Some Ottoman rulers actually partnered with Jews in the wine business, as a way to raise money through taxes and exports. Around 1552 Sultan Suleiman I brought the Jewish House of Mendès into the royal fold, including the Jewish diplomat Don Joseph Nasi. In 1566 Selim II (aka Selim the Sot) became sultan, and he gave Nasi control over a vast system of wine production and export, in Ottoman territories throughout the Aegean, Cyprus, and the Middle East. Historian Avigdor Levy found that Nasi made an estimated fifteen thousand ducats a year from the wine trade. I found records from Cyprus showing that more than 150,000 cases were exported to Venice and London between 1746 and 1770. Parts of the Middle East kept a flourishing export business, even under Ottoman rule. That suggests far more than occasional wine drinking.

The whole situation made me reflect. The original *Oxford Companion* claims that captured my attention, about wine vanishing from Israel and the lack of local grapes, symbolized something bigger about the wine industry: the long tradition of exaggerating the superiority of French wine grapes. Ernest Hemingway coined an insightful word in *A Farewell to Arms*: "I had drunk much wine and afterward coffee and Strega and I explained, *winefully*, how we did not do the things we wanted to do; we never did such things." (Emphasis mine.)

To be honest, my hotel room wine, though real, contained a significant portion of winefullness. I daydreamed about ancient wine before I even knew what it was, and that realization made me almost sympathetic to another case of wine-lover exaggeration. In his popular book *Inventing Wine*, author Paul Lukacs claimed that for much of history wine was valued more for buzz than taste.

"In fact, the descriptions of most ancient wines make them seem unattractive if not downright dreadful. To contemporary tastes, they probably would be virtually unpalatable," wrote Lukacs, a professor of literature. He also believes that the wine lovers "that came of age following the Second World War had the opportunity to experience more particular flavors than ever before in history."

The broad suggestion that all ancient wine would taste dreadful to us today rang hollow, as did the suggestion that an era that featured the uprooting of millions of acres of distinctive regional grape varieties somehow led to a broader range of flavors. Lukacs was making another plug for the superiority of modern French wine. To which I say: Egyptians built pyramids, and the Greeks pioneered math (Pythagoras), astronomy, philosophy, and literature. Yet they couldn't make good wine? I asked McGovern for his take. "I agree with you that pre-1000 AD wine must have had its allurements, since why gush over some of it, age it, etc.?" he replied in an email.

I checked with Rick LaFleur, an emeritus professor of classics at the University of Georgia, who can actually read ancient texts in the original. LaFleur thought the blanket claim of bad ancient wine was wrong. "How can one be a wine snob from a two-thousand-year remove?" he told me. "The ancients loved wine; well, some just drank it, just as some of us just drink PBR; but others, the middle and upper classes, the intelligentsia, the gourmets, and the wine-snobs of their day, went crazy for certain varieties and vintages."

So ancient wine varied, just like people, and just like modern wine. There was good wine, bad wine, cheap wine, expensive wine, ceremonial wine, medicinal wine, spiced wine, new wine, aged wine; in red, white, and rosé; from vineyards that flourished, died, and flourished again as time passed. Pagans, Christians, Jews, and even Muslims drank wine, sometimes secretly, sometimes not.

Thinking about Lukacs I felt a tinge of recognition. As the old saying goes, "There but for the grace of God go I." He envisioned ancient wine one way; I looked at it from another perspective, one that was still essentially an educated guess. Both of us were influenced by memories, individual taste buds, and who knows what else.

TASTINGS

I believe Shivi Drori's research will eventually lead to many new Holy Land wines made from local grapes. Recanati Winery is already making one, and Ido Lewinsohn is experimenting with others. Keep an eye on Yiftah Perets at Binyamina Winery, too.

RECANATI WINERY

Marawi (white, from Hamdani and Jandali grapes)

José Vouillamoz has done new research in Lebanon, too. He found that the local Obaideh grape has a unique genetic profile. Both Chateau Musar (www.chateaumusar.com) and Château St-Thomas (www.closstthomas.com) in the Bekaa Valley make Obaideh wines.

15

American Wine Grapes

I will explain to you the terms by which we characterize
the different qualities of wines . . .
—Thomas Jefferson, 1819

I still had more wine blinders to confront. I had ignored
pretty much anything grape having to do with America. More specifically, I had ignored comments from my
mother and late stepfather over the years, regarding their house in
Vevay, Indiana. Both of them loved historical societies, architec-
ture, American vernacular, that sort of thing. They'd mentioned,
numerous times, that their 1814 home was once owned by a family
that played a key role in American wine history. I always replied
with feigned interest, and promptly forgot the matter.

I had time to read scientific papers about the DNA of wild
yeasts that live symbiotically with wasps, or the particular markings and shape of ancient Egyptian and Roman amphora stoppers,
or ancient Caucasus wine myths—all of those subjects, and dozens
more. But no time to take note of the fact that Vevay was named
after the Swiss town of the same name, and that it was founded by

the man who many considered to be the father of winemaking in America, Jean-Jacques (or John James) Dufour. He was so under my radar that I stayed in the house for a week, helping my mother move after my stepfather's death without ever bothering to investigate the Dufour story. Much later a little voice in my head said, *Oh. Maybe you should look into that.*

Dufour emigrated from Switzerland along with dozens of family members and associates, with the goal of creating the first commercially successful vineyards in America. In 1801 he successfully petitioned Congress for a special land grant on the banks of the Ohio River, predicting that the local vintages would soon rival those from the Rhine and Rhone regions. He sent wine to President Thomas Jefferson, and in 1826 wrote the first winemaking book published in this country, *The American Vine-Dresser's Guide*. At first the Vevay wines were successful and popular, but the vineyards eventually died out, apparently from a combination of disease and neglect.

I started to take the idea of American wine grapes more seriously, even though some see nothing but three hundred years of duds (think Concord grape wine, then banish the thought quickly). But how could I place Saperavi, Himbertscha, Jandali, and all the rest of the rare Old World grapes on a pedestal above every grape in America? When I read about Dufour the implication was clear: my mind had discriminated against the very notion of American wine grapes just as other people turned up their noses at Cremisan or ancient wines in general. I'd cheered when Sean Myles, the grape scientist, spoke about "viticultural apartheid" and the folly of describing a few famous European grapes as "noble." Yet I was just as biased, only in a different way.

But what counts as an heirloom grape if your region has no history of winemaking? Could a quality American wine grape exist? For more than a century the stock answer was no, but that is changing.

George Washington, Thomas Jefferson, and many others experimented with vineyards, mostly with disappointing results. Long before that, Spanish priests brought vine clippings to the New World, hoping for bountiful harvests. So-called Mission grapes flourished in California during the 1800s, but over time were eclipsed by French varieties. Greek, Sicilian, and Swiss immigrants also came here with a love for wine, but weren't impressed by our native grapes. That, at least, was the general story line, and the explanation for why California and the vast majority of American vineyards have planted well-known European grapes for the last seventy-five years.

It's hard to say exactly why native American grapes are so different, but one clue comes from a hundred million years ago. North America was once smooshed together with Europe, South America, Africa, and much of Asia into a supercontinent called Pangaea, which slowly broke apart. After the Americas drifted away our grapes evolved separately, and while there is plenty of diversity here, the flavors they make aren't the same as European or Middle Eastern grapes. Concord grapes make nice jelly, but as the *Wine Grapes* authors politely note, "Wine made from American species, particularly *Vitis labrusca*, can have a very distinctive flavour—definitely an acquired taste—combining animal fur and candied fruits." Others mention aromas that call to mind the time my dog rolled in the mud, after being sprayed by a skunk. America supposedly struck out on native wine grapes.

Or did we? Recently scientists in Minnesota, California, and other states have taken a harder look at indigenous American grapes and found long-hidden qualities. In the early 1980s Minnesota appropriated funds for a grape-breeding program. The goal was to get a promising hybrid into the market. "It really took nearly twenty years to get Frontenac, our first variety, out [into vineyards]. It was actually crossed, I believe, in 1977," said Matthew Clark, an assistant professor of grape breeding and enology at the University of Minnesota.

Botanists crossed indigenous cold-hardy grapes with the European *Vitis*, hoping to create a flavorful, disease tolerant, productive vine. That's more of a challenge than it may seem. "Maybe one out of ten thousand of our seedlings makes it to the stage of becoming a cultivar. So it really is a numbers game," Clark told me in a phone call.

The UM program is doing actual nursery crossbreeding—grape sex—not using laboratory-created genetically modified organisms, often referred to as GMO. I thought hard about the distinction. A decade ago I would have turned my nose up at these wines; not anymore. The new varieties mate indigenous grape DNA with that from European grapes, much like waves of human immigrants changed America. Perhaps you can criticize UM and others for speeding up the breeding process, but so what? At heart they're unlocking flavor, disease-resistance, and growth genes that may be tens of millions of years old. To me these scientists are doing exactly what ancient Babylonians, Egyptians, and Greeks did: refining wine grapes to produce tastes we enjoy.

The UM program is zeroing in on wild and table grapes with distinctive flavors, and working to breed those into wine grapes.

"How much potential do wild American grape vines have, in terms of flavor?" I asked Clark.

"The diversity of flavors is astonishing," he said. "We have grapes that taste like pineapple, strawberry, black pepper. I think the resources are only limited by the amount of time we spend exploring them. We're really trying to develop wine products that are more in the European style, but utilize the resources of the North American germ plasm."

What about the terrible reputation of wild American grapes' flavor, so often described as a kind of damp, mangy foxiness? I realized that must come from a specific—and probably tiny—part of the DNA. Could the nasty aromas be bred out? "We're doing work right now to identify some of the off aromas and flavors, and we're making great strides. Ultimately our goal is to have a DNA test that we can use to screen a seedling years before it produces its first fruit as part of the breeding program, to determine if it has that negative trait or not."

That could be huge, opening up an ancient repository of flavors and aromas that European winemakers never had access to. "The wines we produce, that niche itself, they offer some unique flavor profiles," Clark said. "It's an opportunity for someone who's interested in locally produced products, as well as part of an American story."

A New England winery is showing how much potential the new hybrids have, even if the little town of Woodstock, Vermont, doesn't inspire thoughts of fine wine. There are pottery shops and creameries and white-clad colonial houses near the banks of the Ottauquechee River, and covered bridges and ski resorts in the nearby countryside. So when the *New York Times* listed a vintage from Deirdre Heekin's La Garagista vineyard as one of the top ten wines of 2015, it was roughly akin to an Oklahoma restaurant winning praise for best sushi.

"A few years ago, I never imagined I would fall in love with a Vermont wine," *Times* critic Eric Asimov wrote. "[But her] wines are so soulful that they demanded my attention. I was especially taken with the floral, spicy, lively 2013 Damejeanne." Asimov is one of the most influential critics in the nation, and winemakers around the world dream of making his top ten list. The vineyard location wasn't just the only surprise. The wines Asimov loved used hybrid Marquette red and La Crescent white grapes, both created at the University of Minnesota. Heekin didn't foresee winemaking success either, joking in her book *Libation: A Bitter Alchemy*:

> I wish I could trace my family history back to vocational winemakers from Italy, or France, or even California; a story replete with a derelict chateau, or a sprawling stone farmhouse famous in the village for its perfectly cool cellars, redolent of lime and metal. If only my grandparents or great-grandparents had come through the port of Naples, sleeping on lice-infested mats in the ship's hold, holding tight to their wooden chest of seeds and vines, planting a vineyard in a New World row-house garden once they'd found work and lodging. Such stories, as beguiling as I might find them, are not my story.

I mulled over how a tiny Vermont vineyard that's operated for only a few years had attracted such praise, and whether Heekin's grape varieties foretold America's future vineyards.

Heekin's story does actually have a few European roots. The day after they were married in the early 1990s, Heekin and her husband, Caleb Barber, left to spend a year in Italy, captivated by the Slow Food movement, which started there. They returned home to run a bakery, a farm, a restaurant, and ultimately the winery. Heekin told me by phone that at first they didn't plan a winemaking future. She described life before she started making wine: "I'm the wine director at our restaurant. My specialty is Italian indigenous varietals—rare varietals. I was always predisposed to really looking at what comes out of a region, as opposed to international varietals, like Chardonnay, Merlot, etc." She discovered Coenobium, a white wine made by Italian nuns. The couple visited the nuns' winery, and back in Vermont Heekin had a simple revelation: "The nuns make an elixir that is a manifestation of what nature has to offer them that year. It entices us, intrigues us, makes us thankful, makes us think, makes us remember. This is what the nuns have done." Coenobium is a blend that includes some Malvasia, the same grapes in Leonardo's vineyard.

As the couple learned more and more about their Vermont land, they began to look at American grapes as something more than table fare. "[W]e knew it would be an enormous undertaking," Barber told the *Times*. "Yet it made so much sense. We try to source everything from nearby. If we could have wine from here . . ."

Heekin's visit to another Vermont winery challenged the old narrative that America has only "bad" wine grapes. Lincoln Peak Vineyard was growing some of the new hybrids, and she fell for the wines. Heekin and Barber bought seedlings to plant on their farm, and she sought guidance from other cold winemaking regions. "Because we don't have a history here of what the style of wine is, and what traditional growing is," she told me, "I made a study for myself of every alpine region wine I could get my hands on. Not so much trying to duplicate those wines, but what can I learn from them that gives me inspiration in ways in which we can work with the wine in the field, or styles that we can employ in the cellar in terms of our fermentation." Heekin didn't want to copy wines from, say, Germany—she wanted to learn techniques from winemakers in similar climates.

Heekin said the wines made from the hybrid grapes are exhibiting terroir that is different from the other vineyards they farm. "What we're starting to see, which is super exciting, is that the varietal is there, but its presence is starting to fade, and the site, the presence of the personality of the vineyard, is starting to take over in terms of the impressions you get from the wine," she said. Their wines are a raging success. The couple has now closed the little restaurant, and opened a small wine bar and tavernetta at the vineyard.

Still, much of America is warm, not cold. That's where a quirky, legendary winemaker and the University of California, Davis, come in.

Coyotes, deer, doves, songbirds, and turkey eat Canyon grapes (*Vitis arizonica*); Pueblo and Apache tribes, Spanish missionaries, and early European settlers did, too. The grapes have a natural immunity to many diseases, so they flourish from Texas to Arizona and down into Mexico, in pine forests, riverbanks, and floodplains. Now a romantic, radical, and somewhat perplexing vineyard experiment in California may help winegrowers discover new flavors, and avoid a ruinous and persistent disease without using chemicals, partly by using a portion of *Vitis arizonica* DNA.

"I'm not sure if it is a lightbulb or a colossal delusion," California winemaker Randall Grahm told me, adding that it may take a generation to find out. In an industry in which creating a new logo can qualify as innovation, Grahm is more like Jimi Hendrix playing "The Star-Spangled Banner"—poetry, distortion, and perhaps just a little madness, but art, whether you like it or not. His new California vineyard, Popelouchum, an offshoot of Bonny Doon Vineyard, is a traditional, organic, scientific, pagan, open-source approach to winemaking that just might change the winemaking world. Or not. The 280 acres are about thirty miles southeast of Santa Cruz, set amidst rolling, windy hills.

After decades of fame, fortune, and apparent success as a California winemaker, Grahm decided to try the most taboo yet obvious vineyard experiment of all: let vineyard grapes have sex out in the real world so they create ten thousand new varieties. Philosophically, it boils down to this: he hopes that by marrying the vines to the California soil and climate from birth, nature will release more flavors, or variations of flavors, than controlled breeding ever did. If that seems like an odd plan, it's only because we rarely think about the Faustian bargain at the heart of the wine industry: stopping grapes from reproducing. For decades I never did.

Practically, Grahm is trying to mix the most primal form of wine cultivation—putting seeds in the ground—with modern nursery breeding. Vouillamoz and Myles taught me about the secret at the heart of modern winemaking—lock in the tastes but shut down any evolution, thus creating an opportunity for pandemic diseases to take hold.

I asked Grahm what inspired Popelouchum, and he told me it wasn't so much a specific moment as a realization that came to him around 2012. "It just struck me that if you're going to try and

produce a wine of terroir, you're going to have to use a very different approach."

For years Grahm seemed to have mastered the standard play-
book. In the 1990s he built a national reputation with labels such
as Big House Red and Cardinal Zin. He's won three James Beard
Awards, sold the popular brands to a wine conglomerate in 2006,
and was inducted into the Culinary Institute of America's Vintners
Hall of Fame in 2010. Yet Grahm talks openly about his failures.
"My wines were OK, but was I really doing anything distinctive
or special? The world doesn't need these wines—I was writing
and talking about terroir but I wasn't doing what I was saying," he
told the *New York Times* in 2009. Not only that, Grahm confessed
to using the hidden tools that many successful, modestly priced
wines rely on, asserting that he no longer wanted to use winemaking tricks such as aroma-enhancing yeasts, enzymes, and spinning
cones that artificially lower alcohol content.

Grahm continued the lament with me, and it was like hearing
Eli Zabar admit that his bagels weren't so special after all. Grahm
did more than dis Chardonnay and Merlot. He's famous for worshipping Pinot Noir, an elegant grape used in Burgundy, but now
feels he spent decades laboring under a fundamental delusion.
"It's very funny," he said. "I just made this self-discovery the other
day, and that is: All I ever dream about is Pinot Noir—or I should
say Burgundy. All I want to do is produce Burgundy [wines]. The
problem of course is I can't produce Burgundy in California. For
a long time, I think I thought 'Oh, if I could just grow Pinot Noir
that was really great, and emulate Burgundy, my problems would
be solved.'"

The problems never got solved. Now Grahm feels he made the

same mistake winemakers all over the world have made: expecting French grapes to produce brilliant wines in climates they weren't bred for. In early 2016 Grahm summed it all up by telling a wine critic that California winemakers succeed at control, and that consumers appreciate consistency. "But at the same time, it removes much of the possibility of greatness, or surprise, or originality. And that is the tragic failing of many of the wines of the New World, and not just California."

The question is, can Popelouchum succeed? I called Andy Walker, the wine grape genetics expert at UC Davis, aka the wine academia center of the universe. The school sometimes gets blamed for helping winemakers create generic wines, but Walker's big-picture view echoed Grahm's hopes and concerns. "I do think Randall's experiment is worth doing. Diversity is good. Whether he selects optimal individuals and uses them, or accepts all variation as good and blends the whole mixture is an interesting question and I hope he tests both approaches," Walker said.

Could Popelouchum really produce grapes that express terroir and new flavors, I asked? "Yeah, maybe," Walker said. "He'll be letting the environment sort of sort out those siblings. And some will do better and some will do worse. And it's going to happen over decades—it won't be a quick process." Walker added that it's hard to predict the outcome because there aren't many examples of this kind of breeding, where different seeds from grape clusters are planted. The European and Middle Eastern vineyard industry spread via cuttings, not seeds. The biological details of Popelouchum brought to mind something Vouillamoz had told me: sowing all the seeds from a single bunch of grapes doesn't produce identical vines—each seed will produce a plant with its own flavors, just as a mother's children can vary in hair color, size, eye color, etc. "[He] should get a large range of variation. And that's what he's looking for," Walker confirmed. Grahm might be getting more than he bargained for, but creating just a few new wine grapes would be a tremendous success.

Scientists are getting much closer to understanding exactly why certain grapes or yeasts express particular flavors. Walker said that there's a huge revolution going on right now based on new technology that can sense and detect flavors. "The nose was the ultimate arbiter until very recently. Now we have electronic noses." According to Walker, that creates the possibility of precisely identifying chemical compounds in wine and in grapes. It's not clear where all the powerful new tools will lead, though. "No one has really asked the next question, which is: What do we do with this stuff now?" What he means is whether the industry, and thus wine lovers by extension, will embrace the new wines or stick with traditional options.

There is another selling point to the new hybrid grapes, however. Unlocking the natural disease-resistant genes in indigenous vines could make the new grapes hard to pass up. Pierce's Disease (PD) causes entire vineyards to wither and die. It's transmitted by winged insects called sharpshooters, which are typically a little less than half an inch long. You've probably seen one if you look closely at plants—they're like mini cicada-grasshoppers. A UC Davis study of PD economic costs and impacts found that between 1999 and 2010 the industry, plus federal, state, and local governments have spent nearly $544 million on fighting PD and the sharpshooter pests. Despite those efforts, California vineyards lose about $56 million each year from vines dying from PD. And it's not just California; this is a worldwide problem.

The money is only part of the story. Walker said vineyards apply huge amounts of fungicide to control other grapevine diseases such as downy mildew. The general practice is to spray before the disease even appears, as a safeguard. "It's like saying, 'I'm not going to allow any leafhoppers, so I'm going to kill every one in the whole area before they can possibly get my vineyard.'" Walker understands the pressures vineyards face from potentially devastating loses, but he said the public isn't likely to keep accepting such chemical and pesticide use forever. It's a modern version of what the French went through after phylloxera. They tried soaking vineyards with every chemical imaginable, but the ultimate solution was grafting the European vines on American rootstocks, which had natural disease resistance.

"You can envision a time [when] we have to do something about it. And we can," Walker informed me, by selectively breeding wild grapevine resistance into wine grapes. Walker has created grapes that contain just three percent *Vitis arizonica* DNA, yet they retain the natural PD resistance. Walker thinks we're approaching an era when resistance to diseases, mold, and even drought can be bred into wine grapes. If he's right, we could vastly reduce the amounts of pesticides and chemicals that get dumped on vineyards each year, all of which ultimately end up in the soil or nearby waterways.

It could be a historic development, but science isn't the only limitation. A leading California newspaper used the term "Frankengrapes" to describe Walker's research, even though he does natural, old-fashioned plant breeding. I sighed as Walker told me the story. Our DNA has traces from many other species, too. The "Franken" suggestion probably generated clicks, because activists use the word to describe the type of GMO crops Monsanto sells. The paper eventually changed the headline, and Walker said the reporter didn't aim to denigrate his work. The fact remains, though: a wine grape with three percent wild DNA raised red flags for one writer. Yet consider the supposedly lowly parents of some of the famous French grapes. Nothing is "pure." That's just a wine industry illusion. Would any foodie balk at a splendidly flavorful tomato, just because three percent of its DNA came from natural cross-breeding with some other tomato variety? No. People freaked out at early GMO experiments because scientists mixed genes from different species—for example, flounder DNA went into one tomato variety to help it resist freezing. Many scientists point out that DNA is, well, just DNA—a set of genetic instructions. The flounder-tomato mix helped spawn the Frankenfood term. It's true that American grapes have a reputation of producing iffy or worse wine, but using portions of their genome is a different matter. I asked Walker if that aversion is a purely psychological barrier.

"Yeah," he said. "And in fact we'll have to get over it," given the social pressures to reduce chemical use and the way climate change is already impacting some wine regions. Luckily, Walker said, there's no need for GMO vines, because of all the genetic diversity in wild grapes, plus all the lesser-known native grapes being rediscovered. "There's no reason to use genetic modification unless you don't have the genes at hand. And within *Vitis* we have everything we need."

Walker's work with wild American grapes may deliver many more genetic surprises. Over the last thirty years, he's collected about twelve hundred different samples from across the southwest. DNA analysis by other scientists has also unraveled the true story of California's Mission grape, reputedly brought here by Spanish missionaries in the 1600s. The grape variety turns out to be Listán Prieto, from the Spanish countryside around Madrid. Immortalized in *Don Quixote*, it has some links to Muscat of Alexandria, Egypt. The Mission grapes have brothers and sisters in Peru, Chile, Mexico, and Argentina—all brought to those countries by missionaries between the sixteenth and eighteenth centuries.

For now Grahm has planted some Grenache grapes at Popelouchum, with the goal of breeding disease resistance into existing grapes that already have great flavor. After the first harvests, he'll look for new flavors and try planting other grapes. Popelouchum's Indiegogo campaign raised almost $175,000 and Grahm has secured official nonprofit status for the project. He said they would share the breeding information with other winemakers, and that the first trial wines could be bottled around 2020. A final verdict on whether the idea works won't come for decades, and he thinks even getting to that point will cost "a lot" more money. I suggested that if he created ten distinctive new wine grapes, or even only a few, that would be an achievement.

"It would," he replied. "But it's not necessarily about finding the superior grape for the site." Success could also come from a blend made from many of the new grapes. By early 2018 Graham was trying or considering a wide range of grapes at Popelouchum, including Furmint, Ruche, Rossese, Timorasso, and Ciliegiolo—and still planning the massive grape breeding project.

Grahm's project made me think that saving rare and native grapes won't do enough to change preconceptions about wine. Perhaps it is just part of the effort to encourage new tastes, diversity, and enlightened land use. Grahm summed it up at a Brooklyn food conference, telling the audience, "Why do wines of place matter?

For the same reason that distinct species of butterflies, birds, or salamanders or the discovery of new stars and galaxies matter. They add richness and complexity to our lives."

During our conversation I asked Grahm why he thought wines have stayed so homogeneous in America, given the surging interest in craft beers and whiskies, distinctive chocolates, and foods from all over the world. Did he think vineyards will really change here? Grahm told me he's tremendously encouraged by the younger generation of European wine lovers, who really embrace local wines. He's withholding judgment on our country.

To get a longer perspective I spoke to Kermit Lynch, the legendary wine importer and author of *Adventures on the Wine Route*, who opened his first store in 1972 in Berkeley. Lynch said the irrational craze for wines made from just a few French grapes took off in the early 1980s, when Robert Parker emerged as the most influential wine critic in the country. "My customers would walk in carrying their Parker review," he told me. They wanted whatever the *Wine Advocate* recommended.

Echoing what I'd heard over and over again on my own wine route adventures, Lynch confirmed the same story to be true in California. "Winemakers were pulling out local varieties to plant what they thought were the noble varieties. Well, it didn't quite work out that way. The wrong soil and Cabernet Sauvignon is no longer a noble variety." Wine journalists helped by hyping some French harvest years as the best or worst ever, fueling marketplace swings which promoted the notion that critics could truly discern such trends. Publications started scoring wine on a 100-point scale, and Lynch said that made no sense. "Judging wines like that—that's totally new in the history of wine. I see so many wine experts who know everything about everything." He gently suggested that is unrealistic.

Waves of baby boomers grew up drinking certain wines, and the industry developed tools that place a premium on consistency, not distinctive tastes. Many US wineries sought to copy popular, easy-drinking flavor profiles, with only slight tweaks, like fast food french fries. If a vineyard produced wine with too much or too little sweetness, tannins, or alcohol, there were ways to fix it all.

"There's so many things you can add and take out of wines now," Lynch said. "Somebody sent me a catalog that's sent to winemakers, with all the things they can buy to change the taste of their wine. They can even lower the acidity by adding powder to it. It's just amazing. And of course there must be flavors. I'm sure you can buy flavors." Lynch is right, of course. Do you like that hint of oakiness in certain bargain brands of Chardonnay? Don't assume the wine ever touched a barrel. Some wineries just throw oak chips into stainless steel tanks, and it's perfectly legal to do so.

I learned that the writers Wendell Berry and John McPhee were huge fans of Lynch's classic book, too. Berry, the Kentucky poet and proponent of local, sustainable agriculture, wrote an article for one of Lynch's monthly brochures. "I saw much clearer than I had before how my interest in wine could be accommodated to my interest in good agriculture," Berry wrote about *Adventures on the Wine Route*, calling it "among other things, a fine book on agriculture. One of the best, really, for its interest in the way the quality of place and soil and work are communicated to the quality of the final product. [. . .] Drinking wine from a good little vineyard . . . is like eating vegetables from a fertile, familiar garden, or lamb from the flock of an excellent shepherd whom you know. The immediate pleasure of taste is enlarged and enhanced by the pleasure that one takes in the life of the world and the husbandry of the soil."

My biggest American winemaking surprise came from Oregon, more than eight thousand miles from Tbilisi, Georgia. I never expected to hear about a link to ancient winemaking in the rainy, forested mountains near Portland. Until a few years ago, Andrew and Annedria Beckham didn't either. The couple met in Utah in 1998, while Andrew was studying there, and later moved to Oregon. He got a job teaching ceramics at a school district just outside Portland, and they bought eight acres of land to build a studio and raise a family.

Their neighbors, in their eighties, grew wine grapes, and Andrew thought it would be fun to plant a few rows. Annedria humored him. His teaching job and the pottery studio took up a lot of time, and the couple started a family, so in retrospect a winery was the furthest thing from their minds. But today they make a perfect winemaking team. He has grand dreams; she is a realist. Annedria learned the business side, and became the executive director of the local wine growers association. In the winery Andrew's workday begins at 4:00 a.m. during harvest season. After a full day of teaching he's back at the winery, sometimes until after midnight.

In the early years most of their eight-acre hillside was planted with mostly Pinot Noir–style grapes, plus some Swiss Wädenswil and German Riesling. Their first vintage was 2009. Not so revolutionary—there are plenty of young Oregon winemakers.

Around 2013 Annedria showed her husband an article on Elisabetta Foradori, the amphora winemaker I had tried to visit in Italy.

"I was really enamored with the fact she was using these big clay pots to make her wine. I looked at the photos and said I think I could make something similar. I set out to figure out how to do it," Andrew told me when we spoke on the phone. "It's certainly been a challenge, because there aren't people to ask questions of in the United States. So I've been doing a lot of this stuff through trial and error. I've gone through extensive trials with different firing temperatures and different clay formulations, to come up with a body that's fired at the right temperature, is porous enough for the gas exchange I'm looking for, but is not so porous that we have problems with sanitation, and volume loss, and [wine] leaking through the vessel." A large amphora can take more than two days to fire, and three months to fully dry. They must have integrity and strength after firing, yet in ancient times they were built in enormous numbers, which amazes Andrew.

At first the winery focused on establishing the vines. Traditional fermentation was done in stainless steel and oak barrels. But by 2013 Andrew was ready to experiment. Today some vintages are fermented in stainless steel and aged in amphoras, some use traditional barrel aging, some use amphoras for every step, and others remain in the stainless steel for the whole process. "In the cellar at this point we're able taste how those four styles of winemaking are affecting the same lot of fruit. And it's incredible how different those four wines are," he told me.

The Beckhams farm organically, using indigenous yeasts and bottling most of their wines without filtration or other manipulation, just as everyone did for thousands of years. Andrew is collecting precise analytical data about how the wines made in each type of fermentation vessel age. "The wines that we're aging in clay seem to be evolving at a much faster speed than wines that are aging in wood, so the wines coming from clay are higher toned. There's more energy to them. They all have a very dusty, earth-like texture and intense clarity." I'd noticed that in Georgian wines, too. The clay gives an earthy, iron undertone to the wines, but somehow makes fruits, spices, and other aromas pop out, like clear notes on an acoustic guitar, not a fruit bomb. Andrew thinks that's because the amphoras act as clarifying and refining chambers.

The Beckhams only produced about three thousand bottles of amphora wine in 2014, but they saw a pattern. Critics loved the amphora angle. *Forbes* published an article online, and *Food & Wine* and *Wine Enthusiast* magazines did stories, too. An Oregon wine critic wrote that America finally had truly traditional winemaking—the "Slow Wine" version of Slow Food. At one point Andrew loaned an amphora to a fellow winemaker, and soon people were calling him not just about buying wine—but also about buying amphoras.

He became the first person in America to manufacture huge Caucasus-style clay amphoras. He scaled up to make some that can hold three hundred gallons (each one can require up to 1,500 pounds of clay) and zeroed in on a specific clay from the Sacramento Delta. "Everything has to be just so for the body to work," he explained, from the firing temperature to lining the inside of some amphoras with beeswax. His pottery knowledge—and his kiln—led to the discovery that the amphoras can be re-sterilized after each vintage with a quick burst of heat.

I told Andrew about the scientists who traced many ancient

amphoras to a region near Gaza. "That's not surprising in the least," he said. "And you add the craftsmen to that—people having the technical know-how and the ability to fire. Having a single source for those early vessels would make sense to me."

The story of a young couple growing their own grapes and making their own amphoras drew a lot of attention. Andrew was pleased about how the wine community responded but he also admitted to some frustration at not having enough product to spread their vision: amphora wines that appeal to a broader public. "Some of the wines that are made in clay are very austere. And they're really hard to get your head around. If you're a winemaker, they're kind of fun to geek out on. Our goal is to make really interesting, compelling, well-made wines that people can enjoy and not just want to have a sip of and talk about. The last thing I'm doing is just sealing them up and opening them a year later and calling them good." He hoped more winemakers using his amphoras would feel the same.

The Beckhams had all sorts of responses to the amphora wines. "Reactions from, 'There's no way that's Pinot Gris,' to 'that is the most compelling and interesting Pinot Gris I've ever tasted,' Andrew told me. They've also noticed something Randall Grahm talked about in California. "You have to search hard to find a wine made in Oregon that's done poorly," he said. "But the problem is, they're all so similar. In my opinion what we make here in the [valley] with our Gris is so boring. It's great to drink on a warm day, but has nothing special to say." That goes for many red wines, too. "They're beautiful—but there are so many of them that are the same."

Andrew said he was still exploring why different countries

made the containers in various shapes. He sounded enthusiastic but slightly unsure of where the project was headed. Would he focus on the amphoras, on traditional varieties and aging, or try a whole new variety of wine, like Savagnin? Would he continue to sell amphoras to other wineries? I made a note to check back in, though I didn't expect any dramatic developments. I was mistaken.

I called Andrew about a year later. The amphora business was booming. "We built a huge new production facility and I was fortunate to acquire some equipment that will allow me to take things to the next scale and size. I've got people chomping at the bit for them this year."

An Oregon potter starting an amphora factory? Sure, I could see special orders here and there, but a real business? There was so much interest from other winemakers that the Beckhams saw great potential in being the first commercial amphora maker in America. They bought a kiln so big a car can fit inside, and a crane to handle larger sizes. He's making 500- and 1,000-liter sizes now; a 2,000-liter model is coming soon. Their winery is shifting more towards amphora-style wines, too, and the Beckhams are grateful for the positive response.

Andrew said one of their new amphora wines sold more than half the available vintage in the first six weeks on the market, despite a forty-five-dollar retail price.

A long way from Cremisan, Jerusalem, and the Caucasus Mountains, ancient-style winemaking was growing in America, so I asked Beckham a question. "This might be crazy," I said, "but would you be interested in trying to re-create an ancient-style wine?" I told Beckham about my travels and mentioned Patrick McGovern's work on ancient ales with Dogfish Head Brewery. He knew, of course, that, amphoras would be crucial to making a wine that is something like what the ancients drank. I didn't really expect Beckham to agree—he's got the full-time teaching job, a wife and three young children, vineyards, a winery, and a new qvevri factory. I finished talking and there was a brief pause on the end of the line.

"I'd be really interested in that," Beckham told me, and I jumped up out of my chair. We discussed some grape possibilities, and agreed to keep in touch.

Oregon may be the epicenter of wine innovation in America, even if there's more money in California or New York. Chad Stock of Minimus Wines in Portland pushes winemaking traditions by experimenting in every vintage with different grape varieties, yeasts, fermentation, or aging. Here's how he described their 2015 SM3 vintage: "SM3 is always pure Syrah from the Stella Maris Vineyard (SM). Fermentation is 100 percent wholecluster with maceration lasting one month. Daily foot treading is used to manage the cap for gentle extraction. Aging is done in 33 percent new, and 67 percent used French oak for ten months before being bottled unfined and unfiltered without any added sulfites ever."

The translation of all that is Minimus mostly lets wines take their own course. Fining is a common industry process that can use a long list of compounds (fish bladder extract, seaweed, egg whites, clay, and more) to remove tiny suspended particles from wine before bottling, making it clearer. Filtering does the same thing.

Another Oregon producer helped me understand a piece of historical writing that had baffled me. Pliny the Elder compiled a bizarre list of so-called wines two thousand years ago, and I couldn't imagine what any of them tasted like:

A wine is made [from] the pods of the Syrian carob, of pears, and of all kinds of apples. That known as "rhoites" is made from pomegranates, and other varieties are prepared from cornels, medlars, sorb apples, dried mulberries, and pine-nuts; these last are left to steep in must, and are then pressed; the others produce a sweet liquor of themselves. [. . .] Among the garden plants we find wines made of the following kinds: the radish, asparagus, cunila, origanum, parsley-seed, abrotonum, wild mint, rue, catmint, wild thyme, and horehound. A couple of handfuls of these ingredients are put into a cadus of must, as also one sextarius of sapa, and half a sextarius of sea-water. [. . .] Among flowers, that of the rose furnishes a wine.

That's just a partial list of strange "wines" that Pliny mentions, and it all suggests an attitude of "if it can be fermented, we'll drink it." But how would any of them taste? I never expected to find out, until I learned of Carla David's Wild Wines in Jacksonville, Oregon, just north of the border with California. In 2007 she started making wildflower and fruit wines out of her garage. Now, after receiving a US Department of Agriculture grant, she is producing more than three thousand bottles per year. Among them are dandelion, peach, ginger, linden flower, elderflower, raspberry, blackberry, elderberry, and rosehip wines. I ordered a few bottles, half expecting sickly sweet concoctions.

The Elderflower conjured up images of a pollen-covered bee. It was crisp, dry, and more floral than anything I've ever drunk. Is it wine? Technically, perhaps not. She starts with water and adds yeast, flowers or berries, and some sugar. Here's the weird thing: in a blind test I bet some of her wines could fool people into thinking they were made with grapes. That challenged my preconceptions again.

TASTINGS

I've mentioned some special wineries in America, but many others are experimenting with lesser-known wine grapes and American–French hybrids. To get a sense of the winemaking diversity in Oregon, check out *Voodoo Vintners* by Katherine Cole.

BECKHAM ESTATE VINEYARD, SHERWOOD, OR

beckhamestatevineyard.com

Beckham combines fermentation in amphoras with aging in amphoras and oak barrels, and has a large pottery studio, too. In addition to the wines listed below, they made a very nice amphora Malbec in the past; keep an eye out in case they make another in the future.

A.D. Beckham Amphora Pinot Noir "Lignum"

A.D. Beckham Grenache (red)

A.D. Beckham Pinot Gris

Vermentino (white)

MINIMUS WINES, CARLTON, OR

www.minimuswines.com

Rockwell (a blend of red and white wines)

SM2 (a blend of Viognier and Sauvignon Blanc grapes)

WILD WINES, APPLEGATE VALLEY, OR

www.enjoywildwines.com

Wines to try from Carla David's venture include Aronia Berry, Lindenflower, Elderflower, Rosehip, and Dandelion.

LINCOLN PEAK VINEYARD, NEW HAVEN, VT

lincolnpeakvineyard.com/wines

This is the winery that inspired Deidre Heekin to try the new American–European grape varieties. It's hard to keep up with all their innovations: Lincoln now has a new variety, Petite Pearl, and also uses local Marquette, Louise Swenson, La Crescent, Prairie Star, and Adalmiina grapes.

LA GARAGISTA, BARNARD, VT

www.lagaragista.com

Many of their wines sell out quickly, but here is a tasty solution: Deidre Heekin and her husband, Caleb Barber, opened a rustic wine bar and tavernetta at the vineyard in early 2017. Visit and try vintages such as Damejeanne, Harlots and Ruffians, Loups Garoux, Grace and Favour, and who knows what else, along with prosciutto, fruits, and vegetables from farms in the region.

POPELOUCHUM, SAN JUAN BAUTISTA, CA

www.facebook.com/Popelouchum

Randall Grahm may have some small batches of wine from his experimental vineyards in 2018 or 2019, check the vineyard's Facebook site periodically for updates.

CHANNING DAUGHTERS, BRIDGEHAMPTON, NY

www.channingdaughters.com

This Long Island winery make a variety of "orange" wines in the Georgian style, including one that uses the ancient Ribolla Gialla grape.

DR. KONSTANTIN FRANK WINE CELLARS, HAMMONDSPORT, FINGER LAKES, NY

www.drfrankwines.com

Ukrainian immigrant Konstantin Frank (1899–1985) pioneered the introduction of Georgian Rkatsiteli grapes to America more than fifty years ago, and the winery's version is an easy, under-twenty-dollar way to experience that wine. A family enterprise, his descendants run the winery now. They also make some Saperavi reds.

Rkatsiteli (white)

16

The Dark Side of Wine Science

There is more philosophy in a bottle of wine
than all the books.
—LOUIS PASTEUR

The scientists I met throughout my journey were an invigorating lot, full of ideals and energy about wine old and new. McGovern, Vouillamoz, and others used high-tech to illuminate the past, save rare grapes, and tweak winemaking traditions. However, I should have remembered that science can be a wrecking-ball for tradition, too. Outside of the lab it doesn't always go where you expect. Einstein changed our understanding of the universe, which helped lay the theoretical foundations for atomic weapons. DARPA, the federal military research agency, created the proto-Internet; today we have (so far) Google, Facebook, and Twitter, which seem to rule many people's lives.

Take Ava Winery. Started in San Francisco by two recent biotech grads, Ava aims to create synthetic wines without using grapes. The end product is about 85 percent water, 13 percent ethanol, and 2 percent chemical flavor and aroma compounds. While I looked to the past for inspiration, Ava looked into the future. "Picture how humanity creates food in 5,000 years. We believe foods will be designed the way clothing is designed and printed with the ease that we print on paper today," Ava proclaims in their mission statement. "The future of creating foods with total control can't be realized if we don't understand the 'inks' that will be used to print these foods. Ava's pursuit of the molecular reconstruction of food will help push the envelope of the food tech revolution."

The pitch brought to mind Star Trek-y gadgets that dispense food and drink with a few buttons, and my appetite was not whetted. Yet Ava was inspired by an experience I could relate to. The founders had seen a 1973 bottle of Chateau Montelena Chardonnay, which beat out French wines in the legendary (to California winemakers) or infamous (to the French) Judgment of Paris, a 1976 showdown between early California wines and Old World ones that wasn't supposed to be a contest, and certainly wasn't supposed to see the French lose (the Smithsonian even has a bottle of the Montelena). When Ava's founders saw a bottle of the wine with a price of over ten thousand dollars they realized it "was hopelessly out of our reach. But what if we could re-create it, molecule by molecule? Certainly, it should taste the same. And while the replica might not bring as much pleasure to our egos as drinking a ten thousand dollar bottle of wine, it should bring pleasure to our palates."

Most serious wine lovers can relate. Yet Ava hasn't taken the industry by storm. They seem to have made the mistake of offering an early batch of the synthetic wine to a few journalists. *New Scientist* compared it to Italian Ruffino, and found, "The smell was the first thing that gave the synthetic stuff away." One person described it as "the smell of those inflatable sharks you take to the pool." But I wouldn't count Ava out yet. Another reviewer wrote that it "tasted better than it smelled," with fruity pear, peach, and even floral notes, and reported that the "winery" was still tinkering with formulas.

Synthetic wines could be more acceptable to future generations. In late 2016 a BBC Travel story featured Gïk Blue, a Spanish company that added pigments to red and white grapes, creating a "sweet, electric-blue wine that has some raising their eyebrows and others raising their glasses." Co-founder Aritz López said he created the wine "for fun," and that he wanted "to shake things up and see what happens . . . and the wine industry looked like the perfect place to start."

Seen from a broader food-world perspective, these shudder-worthy creations may be a trend. Around the same time Gïk released its wine, *New York Times* food critic Daniel Duane reflected, "For baby boomers who moved to the Bay Area in search of the unfussy good life, in the late 20th century, it was all about squinting just right to make our dry coastal hills look like Provence—per the instructions of the Francophile chefs Jeremiah Tower and Alice Waters of the legendary Berkeley restaurant Chez Panisse. [. . .] Today, Northern California has been taken over by a tech-boom generation with vastly more money and a taste for the existential pleasures of problem solving."

Another new product announcement floored me, too. In late 2016 a company called Vinome attracted venture capital with this pitch: "What if wines could be scientifically selected for you based on your DNA?" I had struggled to find obscure wines or afford famous ones, and now Vinome said all I had to do was send them a sample of my DNA to find happiness in a glass. They claimed to have "curated hundreds of scientific studies to isolate genetic varia-tions that are shown to be associated with taste and smell," then arranged for hundreds of people to taste and rate wines, while also testing their DNA for key genetic markers.

"Explore and enjoy! No guesswork. No intimidation. Just great wines, perfectly paired to you," the pitch said. I was somewhat ap-palled, but still curious. I took their brief test, which featured a half-dozen questions, such as "Do you think you prefer sweet and fruity wines or wines with oak and earthy flavors?" At the end they asked me to agree to the Terms and Conditions, which promised that my DNA would remain mine, except that "By submitting DNA to Vinome, you grant Vinome a perpetual, royalty-free, world-wide, transferable license to use your de-identified DNA, and to use, host, sublicense and distribute the anonymous resulting analysis to the extent and in the form or context we deem appropriate on or through any media or medium and with any technology or devices now known or hereafter developed or discovered."

All of a sudden, my years spent tracking down Cremisan and trips to the mountains of Georgia felt like a reasonable price to pay for a fine wine, compared to forking over DNA to a company I knew little about. I did more research and found that Vinome had partnered with Helix, a global genomics powerhouse. The Helix press release about Vinome opened by touting new partnerships with *National Geographic* and New York's Mount Sinai Hospital. Helix said it is also building new applications "that harness DNA and the science of taste to match personalized epicurean experi-ences and products with consumers."

The health and medicine news agency STAT asked geneticist

Jim Evans, a professor at the University of North Carolina, to comment on Vinome. He said, "It's just completely silly. Their motto of 'A little science and a lot of fun' would be more accurately put as 'No science and a lot of fun.' I'd put this in the same category as DNA matching to find your soulmate. We just simply don't know enough about the genetics of taste to do this on any accurate basis." Not only that, Vinome reportedly plans to sell the wines for sixty-five dollars a bottle. For that price you can find many outstanding vintages at any decent store.

Stepping back, Vinome and Ava made me contemplate how wine lovers got to such a place, where deciding what pleases us supposedly requires a DNA test. Yet perhaps the predicament isn't fundamentally new. Wine drinkers have always been wary of strange mixtures. Steven Shapin, a History of Science professor at Harvard, noted that a character in the 1771 book *The Expedition of Humphry Clinker* laments, "What passes for wine among us [the English], is not the juice of the grape. It is an adulterous mixture, brewed up of nauseous ingredients, by dunces, who are bunglers in the art of poison-making."

Whatever the future holds for synthetic wine and other creations, regulators are using scientific advances to keep up. Today the US Department of the Treasury operates a Beverage Alcohol Laboratory in Beltsville, Maryland, with a dizzying list of responsibilities: screening products for contaminants, adulterants, and unauthorized additives, as well as handling mislabeling and fraud investigations, smuggling and counterfeiting, and pre-import testing of foreign wines. While alcohol taxes date to the very beginning of the country (they helped pay off Revolutionary War debt) the Treasury now uses liquid chromatography, mass spectrometry, and other tools. Government lab spec sheets reveal that ethanol has a basic chemical formula of C_2H_6O and a molar or atomic weight of 46.0684. That is more than carbon (12.0107) or sulfur (32.065), but less than copper (63.546) or uranium (238.0289).

Perhaps all is not gloom and doom. Honestly made, distinctive wines have found their place in America. For more than forty years Kermit Lynch has sold only wines from small producers. His business has steadily increased. Yes, he told me, there are a lot of easy drinking, not very distinctive wines. But that's part of life. "I'm kind of philosophic about it. It's like in music, pop music. I think there are pop wines—wines made to be popular. That's the idea, the winemaker sets out to do it. And then you have more serious composers."

As much as I love traditional winemaking, vineyards may be forced to embrace some technological change. When my travels began climate wasn't part of the story. Now I see that it is. José Vouillamoz talks about it as he promotes native grapes, even if some consider the subject taboo. "Global warming, or climate change, is a fact. I'm not discussing here the causes—it's a fact," he told me. "And many different areas of the economy will have to deal with it, including viticulture. [In] 2014, at the Masters of Wine Symposium, when I tackled this topic, many people came to me and said, 'Wow, you are courageous. No one dares speak about that.' And I had a fake bottle of Romanée-Conti—I say this as a joke, because it's one of the most counterfeited wines in the world, as well as the most expensive. I faked it on Photoshop; it was obvious. And I put on the label, the vintage, 2214. And I was asking the audience what do you think will be in this bottle, in two hundred years from now. Will there still be Pinot Noir, as it is today, or something else? What are the solutions to face global warming?"

Vouillamoz said Pinot Noir grapes in Burgundy are already out of the optimal window of cultivation because of increasing heat, yet Romanée-Conti's legendary owners would turn in their graves if future generations plant some other variety. It would be like planting date palms to replace the Washington, DC, cherry trees.

"So if you want to keep Pinot you can do adjustments, but at some point you will need some more help," Vouillamoz said. That means somehow tweaking the Pinot variety, perhaps with heat-resistant genes from some obscure ancient vine.

In California Randall Grahm hopes that Popelouchum leads to flavorful grapes that can thrive in a warmer world. I asked him if many other people there are concerned about climate change threats. "The customers and the wine writers are more interested than the grape growers. I don't know that a lot of [vineyards] are thinking about it so much," he said. "Grape growers and winemakers are mostly worried about sales and relationships with their bankers, and wholesalers, than they are grapes of the future. In this country we don't really think much long-term."

I know it's easier to just enjoy wonderful wines, and not think about climate change or pesticides. Just remember that vineyards do impact the environment. The wine choices we make can help lessen the harm.

17

Grapetionary

[It] was like discovering a wine-cellar filled with bottles of
amazing wine of a kind and flavour never tasted before.
—J. R. R. TOLKIEN, LETTER TO W. H. AUDEN, 1955

It is hard to go wrong traveling to vineyards or talking to
winemakers. But can winemakers alone transform America's wine culture? Maybe not. That's why people such as
Jason Tesauro are so important. He is a writer and sommelier who
worked for years at Virginia's Barboursville Vineyards, just over
an hour outside Washington, DC. They're known for fine vintages
made from classic French grapes, and also for successfully planting
varieties such as Nebbiolo, Viognier, Vermentino, Cabernet Franc,
and Malvasia (the grape from Leonardo's vineyard).

Tesauro created an event called Grapetionary A–Z, which gives
American wine lovers a chance to try twenty-six rare, unusual
wines in a single magnificent tasting. Like me, he fell in love with
rare grapes almost by accident, tasting some memorable wines
from Moldova. He puzzled over why more people weren't talking
about them. "I wasn't drawn to 'Ooh, let's take a trip to Napa. Ooh,

let's take a trip to Bordeaux.' I was like, wait a minute. They're growing grapes in Moldova, and they've been doing it for hundreds of years? How have I not tasted this? And so I jumped on a plane."

His feature article about the trip won a national journalism award. Looking back on his early years as a wine lover in the 1990s, Tesauro saw what I did: a wine world featuring just six or seven grapes. Anything outside of the classics was unusual. Being in Moldova changed how Tesauro thought about wine. "The more I drank [those wines], the more I wanted to connect with, 'How did they get lost?' That question brought me to the geopolitical stories of how the Russians kept the Moldovans under their thumbs, and this is why I got into wine, for something more than just the transference of alcohol into my system. Even more than the flavor, it was the passport to another time, another culture, another place. And I found that those indigenous grapes, in particular, were the fastest way there. Because instead of us talking about a winemaker's ego, you know, 'what does he do or she do with new French barriques and micro-oxygenation?' it was, 'who were the stewards of these grapes that were traditionally grown in this one spot?'"

"Part of what I love about wine is that egalitarian side," he added. "If you spend your life running around saying, 'Oh, I've been to the great chateaux and I've had all the premier cru,' then you're kind of nosing those who were behind you. Whereas going and visiting places that are not yet on the map, so to speak, you're back to that place of curiosity, of teach me, oh wise one, tell me what I don't know, and don't just let me spout off about what I do know."

In late 2015 Tesauro developed the first Grapetionary, pitching the idea to the Atlanta Food & Wine Festival. Their initial response was "Are you out of your mind?" But the organizers quickly came to love the idea. Logistics and public response were the biggest questions. They needed a grape variety for each letter of the alphabet, which meant sourcing vintages from Lebanon, Turkey, Georgia, Austria, and many other countries. They also needed ten bottles each of all twenty-six wines, to handle the target audience of a hundred subscribers, each paying $275 for an experience, which also included exceptional food. The response was overwhelmingly positive. A *Forbes* reviewer called it "the most unique wine event I've ever experienced." Well-known people in the wine world took notice. Randall Grahm, Jancis Robinson, and José Vouillamoz all contacted or praised Tesauro.

Tesauro held another Grapetionary in Washington, DC, with food from Michelin star chef Nicholas Stefanelli, and then another at a private charity event organized by a Virginia billionaire. A *Washington Post* headline read, "From aglianico to zibibbo, he poured an alphabetical tour of wine grapes," and the reviewer wrote, "Even for a jaded wine fiend like me, there were some surprises." Tesauro said people at all the events become kids again. They have fun trying to pronounce the strange names, and he loves that playful aspect. "All the wines are good, some of the wines are great. Forget whether or not the wine tasted of blueberry or strawberry. Did it move you? And how did it move you? Does it conjure that last great trip to Greece? My role is simply to play Johnny Grapeseed, and say, 'Did you know . . . that Yapincak is thriving in Turkey?'"

The unusual grapes are a learning experience, too. Take Yapincak for example. Turkey's eastern vineyards are in the area where Noah's Ark reputedly came to rest, which is also part of the broad

region that McGovern, Vouillamoz, and other scientists have identified as the likely birthplace of winemaking. "I had no idea this
grape existed," Tesauro said of Yapincak. "It is now offered by the
glass in the best wine bar in Richmond, because I brought a bottle
to the proprietress and said, 'I know you dig off-the-wall stuff. This
wine is going to knock your socks off.' Now Yapincak sales have
taken off in the mid-Atlantic." It all happened through word of
mouth from sommeliers, restaurant owners, and customers. "Yes,
it's so easy for people to just reach for what they know. But you give
them a chance to taste something that's completely new—but has
an old, earthy, and authentic story, with real terroir and real flavor.
They will recognize the trueness of that, and it will strike them."

Listening to Tesauro, I thought back to the time eight years earlier when people gave me strange or puzzled looks about Cremisan,
or grapes such as Jandali. My point? A wave of diverse, flavorful
wines are more and more available all over the country. You don't
have to go to the Caucasus Mountains to try them. For a parallel,
look at what has happened with food.

In the early 1980s I used to go fly-fishing in the Catskill Mountains in southeastern New York. One day I noticed a bakery in a
ramshackle building along the side of the road, near the teensy
town of Boiceville, in the middle of nowhere. The bread was magnificent. Huge, round, crusty loves of chewy white, hearty raisin
walnut, and other specialties. Bread Alone Bakery now has four locations in the state, and founders Daniel Leader and Sharon Burns-
Leader have a son who joined the business.

At the end of that decade I spent part of the year working on
a Hudson Valley farm that made goat cheese. One milking shift
started at four in the morning, another at five. Over time I learned

how to make the basic cheeses, taking the Sunday morning shift in the creamery. The place was Coach Farm, and in 2017 they're still going strong. Wine today is where local foods were twenty years ago. Change is coming.

TASTINGS

Here are a few of the wines (and grapes) Jason Tesauro has surprised people with so far:

Agiorgitiko, Aivalis Winery, Greece

Emir, Turasan Winery, Turkey

Yapincak, Pasaeli Winery, Turkey

Kisi, Schuchmann Wines, Georgia

Obaideh, Chateau Musar, Lebanon

Ucelùt, Emilio Bulfon, Italy

Blaufränkisch, Weingut Glatzer, Austria

Zweigelt, Weingut Markus Huber, Austria

Petit Manseng, Michael Shaps Wineworks, United States

18

The First Grapes,
and Tasting the Past

I am lost in this city and can no longer find the Winehouse door.
Please help me to find that street again where Love resides.
—HAFIZ, "RIVER OF WINE," CA. 1350

Nine years had passed since my hotel room taste of Cremisan wine. It was mid-January. I was in Massachusetts to speak at Harvard about a different topic, and on that drizzly and chilly day I wasn't thinking about the origins of grapes or delicious, little-known bottles of wine. I met a group of friends for dinner at Oleana, a Middle Eastern restaurant run by chef Ana Sortun, who has won numerous accolades for her subtle yet startlingly flavorful cooking, including a James Beard Award.

The wine list included a bottle of Greek Argyros Assyrtiko, and I pushed our table to try it, keeping my fingers crossed. There were safer choices, and one never knows how people will respond to something new. But everyone loved the bracingly crisp white, which perfectly complimented the lamb and grape leaf tart, celery root dumplings, and other dishes. I was happy to see diversity in the wine by the glass list, too: Vermentino and Cortese whites from Italy, a Hungarian Furmint, and then the reds: a Touriga from Portugal, a Refosco-Mavrodaphne from Greece, a Perricone, a Raboso blend, and an Italian Freisa.

I had a little time before flying home the next day and stopped at Harvard's Sackler Museum, not caring or checking what exhibits were up. It's a small museum with top-notch holdings, a reliable way to spend a pleasurable hour or two. I wandered into a room with some of the best painted Greek and Roman wine vessels I'd ever seen. There was a large krater, or wine container, from circa 500 BC decorated with a procession of bearded men; they traipsed around, carrying amphoras, making music, riding horses, talking to one another. A plate from 480 BC featured a woman playing a drinking game called *kottabos*, which called for flinging the dregs from your bowl of wine at a target. I could see the attraction.

One drinking cup depicted a bare-breasted young woman with what looked like castanets; another had an old woman welcoming an old man—or perhaps telling him to stuff it. A large water jar featured Dionysus, the Greek god of wine, surrounded by a fantastical entourage: a beautiful young woman petting a panther, a bearded satyr, a winged boy flying through the air casting some sort of spell, reclining ladies, and swirling floral motifs on the reverse side. Perhaps they represented vines.

All the rituals and places I'd visited in the Mediterranean were coming to life again, including observations on human nature that are still eerily relevant. One of the gallery cases mentioned a play

the Greek poet Eubulus wrote in the fourth century BC, in which
Dionysus describes the stages of a drinking party:

> Three bowls of wine only do I mix for the sensible: one is
> dedicated to health (and they drink it first), the second to
> love and pleasure, the third to sleep—when this is drunk up
> wise guests go home. The fourth krater is ours no longer but
> belongs to hybris (outrage), the fifth to arguments, the sixth
> to drunken revel, the seventh to black eyes, the eighth is the
> bailiff's, the ninth belongs to bitter anger, and the tenth to
> madness that makes people throw things.

Statues followed the wine vessels as I wandered through the
museum. A nude of Aphrodite adjusting her sandal, from the first
or second century AD; a Roman head of a nymph or hermaphrodite, with a short curly hairdo almost like a flapper; a bronze
of Aphrodite holding a dove, from the fifth century BC. They all
reminded me of her temple on Cyprus, and those island wines.
Two circular busts showed wealthy traders from Mari, the ancient
wine-trading city along the Tigris River. Their neatly coiffed hair
and stern, self-satisfied gazes had exactly the same look you see in
paintings of nineteenth century French bourgeoisie. Prosperous,
and proud of it.

My wine travels had changed the way I viewed history. I first
visited New York's Metropolitan Museum of Art when I was nine
or ten years old, and many others followed: Paris, London, Tel Aviv,
and so on. Yet for all those years I never really saw the *people* in the
ancient stories. They were there, but kind of like elaborate cut-outs.

The museum visit reminded me of a recent *New Yorker* article about star chef Yotam Ottolenghi and a special dinner he gave at the Met featuring Middle Eastern recipes from the medieval era. You may remember it was his sommelier who suggested that José Vouillamoz and his colleagues take note of the Cremisan wine. It turns out that when Ottolenghi was a child, his father took him to the monastery for a memorable tour of the winery.

I reached out to him and asked for a chef's view of the native grapes. Why do they matter, from Cremisan or anyone else? "Well, of course, native grapes are extremely important to preserve because they ensure that we keep enjoying varieties that evolved over millennia," he replied in an email. "Streamlining the production of all crops has been gradually robbing us of diversity of flavour and homogenized our eating and drinking experiences. My very unscientific opinion is that wines made with crops that evolved over time in a particular setting are more interesting, more complex and make much more sense when served with the local food."

In fact, his opinion melded perfectly with those of scientists who have explored the same subject: save rare grapes, enjoy their unique flavors, and you get a biodiversity bonus, too. When grape varieties go extinct we lose those tastes and who knows what else.

Let me confess something. When I set out on all these travels I wasn't sure what I would learn. I'd imagined something about ancient wine, but a part of me debated if the quest had a legitimate purpose. I could have taken sommelier classes instead.

Let me confess something else. Just as I ignored the wine history of my own mother's house in Indiana, I had a scientific blind spot, too. I neglected the man who discovered the world's oldest fossil from the grape family. Oh, I read the research paper, don't get me wrong. But I didn't make the mere four-hour drive from my home to visit Steven Manchester, a paleobotanist (aka a scientist who studies plant fossils) and professor at the University of Florida in Gainesville.

Perhaps my tardiness fit a pattern of roundaboutness. My Cremisan wine obsession started by accident, and I spent years searching for answers. So I finally visited Manchester, whether late or right on time. Seeing his office, I liked him right away. It was small. The desk was crowded. A coffeepot on one side looked as if it could hold fresh brew or weeks-old dregs. The carpet and filing cabinets were kind of worn, but that didn't matter. Manchester focused on his work.

He said grape paleobotany was really in a mess twenty years ago, because scientists didn't completely understand the relationship between modern species and the fossil record. For example, DNA research indicated *Vitis* might have first emerged eighty or ninety million years ago, when dinosaurs roamed the earth. The early grapes split off from the poplar family. Yet the oldest fossils anyone had found—in both southern England and North America—were about fifty million years old. "It seemed odd that we weren't finding fossils as old as we thought we should for the grape family," he told me. Manchester decided everyone might be looking too much in the northern hemisphere, so he sought grant funding for a dig in India. "There were two surprises. One is that [the trip] was funded, and secondly, we actually did find grapes. The molecular evidence is telling us we need to look more in places like India, Africa, Australia, and older rocks," he said.

India was an island sixty-five million years ago, located down near today's equator. An enormous asteroid had hit Earth about a million years before, throwing up megatons of dust and debris that severely reduced sunlight. Many plants died out, then the plant-eating dinosaurs died, and so on. Volcanoes probably helped drive the mass extinctions, too. In India a long range of them were erupting regularly around the same time the asteroid hit, adding even more soot to the air.

"There were huge lava flows coming out. The plants were colonizing those beds, or the soils that develop between those lava flows, which probably wasn't very good for drainage or for some of the nutrients that they needed. So the trees that we're finding petrified there are all small. We're not finding huge ones," Manchester said.

I asked if it was easy to identify fossil grape seeds. "Oh yeah," he said, opening a book and showing me pictures. "So we look for that pair of infolds on one side, and if you turn it over there's either a circular depression or an elongated groove." I wondered what purpose the distinctive markings served, in evolutionary terms.

"We don't know! It probably increases the surface area [of the seed]," he said passionately, then paused. "Maybe *you* know what those extra infolds would do?" he asked. I'm pretty sure he was joking, but his tone was inquisitive, mixed with hope, doubt, and wonder. Anyway, I didn't know.

Manchester's team eventually found sixty-six-million-year-old grape fossils at the India dig, which is fifteen million years older than anything previously found. They aren't exactly *Vitis vinifera*—more like a close cousin. Forty-five to fifty million years ago there was more grape diversity in the world than there is today, as we've lost many species over time, Manchester told me.

I mentioned the widely accepted theory that grapes originally developed flavors to attract animals and birds, so they would eat the seeds and spread them. He suggested it's not that simple, since some of the oldest grapes had nasty, almost trickster-like fruits. A forty-seven-million-year-old fossil had a big quarter-inch seed, with just a millimeter of fruit covering. "Probably wouldn't have been a favorite pick for human domestication," he speculated, and perhaps not even for birds. That could explain the origins of the less-than-appealing flavors of some wild American grapes, I thought. "Some of them have pretty bad raphide crystals in them, probably to dissuade critters from eating them," he added. Raphides are hard, needle-like, naturally occurring deposits found in the cells of some plants.

We talked for a few minutes more, and Manchester offered to show me the oldest grape fossil in the world. In another room filled with boxes and equipment he pulled out a small, clear envelope, and there it was. The thumbnail-sized fossil was encased in black, shiny basalt from the primordial lava flows, and the seed was immediately recognizable, even to me. I took some pictures and asked to hold it. Manchester let me.

I felt a chill or a thrill run up my spine. The little vines must have been real fighters, I thought. Lousy soil, fetid air, and all sorts of hungry creatures trying to eat them. Such a long road to even reach the era about eight thousand or so years ago, when humans figured out how to make wine. It was funny that I'd pegged the six-thousand-year-old Armenian cave winery that Vouillamoz told me about as old, when of course it isn't, geologically speaking.

I still think about Cremisan sometimes, and where those varieties fit on the great wine grape family tree. Now I may even have a reputation among wine scientists as the guy who knows something about Cremisan. Not long after the museum visit, Patrick

McGovern sent me an email asking if Shivi Drori had published his paper yet. Now let's be clear: McGovern, the world's greatest expert on ancient wine, didn't actually ask my opinion. But I confess to kind of feeling like I had arrived.

In early 2017 Drori's team published its research on Holy Land grapes in the journal *Nature*. They confirmed that some of the Cremisan grapes date to the Roman era, documented numerous previously unknown varieties all across Israel, and gave the most detailed picture yet of wine grapes in the region. Drori found genetic links between the Israeli/Palestinian grapes and others in the Caucasus, Iran, and Central Asia. To me the DNA results reinforced previous research and thousands of years of poetry, supporting the theory that humans traded grapes all over the region, seeking better flavors for eating or for wine. Some grapes from the Central Asian *sativa* grape subspecies also have the potential to make fine wine, according to the research, and in retrospect that makes sense. Europeans manufactured the idea that only certain grapes make fine wine. It now seems likely that the first winemakers, from the Caucasus down to the Holy Land, used other grape varieties, too.

Drori's research creates a foundation for a rebirth of native, ancient, Middle Eastern wines, so you may someday find bottles of Ein-Misla, Sorek, Nitzan, and Yael wine. We'll see if Israeli consumers are ready for a change. I hope so—I want to taste all those new, old varieties. But Drori didn't prove that winemaking originated in the Levant, as he'd hoped. In fact, Patrick McGovern and a group of scientists found even more evidence for the Caucasus region, with a 2017 paper about a large jar decorated with grape clusters that was used near Tbilisi eight thousand years ago.

Last I heard, José Vouillamoz had identified a previously un-known grape found only in Lebanon. He never said that decipher-ing the entire family tree of wine grapes would be quick or easy. Perhaps some hidden valley in the Caucasus really is full of pri-mordial vines, like the apple variety in Kazakhstan that scientists have identified as an early ancestor for almost all cultivated ones. Traders and travelers along the old Silk Road carried *Malus siev-ersii* seeds to the east and west, starting at least ten thousand years ago. Their horses even helped by eating the apples and pooping the seeds out later—a package of natural fertilizer. Adrian New-ton, a scientist who has studied the apples, found that a single wild Kazakh *M. sieversii* tree can have more variation in fruit size, shape, color, and flavor than all the species of cultivated apples in Britain. They can be yellowish, purplish, bright green, or a kaleidoscope combination, and vary from plum-sized to what we consider a normal apple size. They are the apple version of what José Vouillamoz dreams of finding—a single ancient grape variety whose DNA shows a link to almost all the vineyards of the world. The "Mother" wine grape.

I haven't seen a grape like that yet, but maybe someday. Then again, did I already taste the past with Cremisan's wines, or Alaverdi's?

Perhaps. Who can say?

ACKNOWLEDGMENTS

THIS BOOK BEGAN almost by accident, but my hotel room encounter with a bottle of Cremisan wine would never have happened without Jessica Gorman and Gwen Darien, who sent me to Amman, Jordan, on an unrelated science writing assignment.

Over the last eight years many people in many countries have helped me understand wine. Special thanks go to all the scientists who patiently listened to my seemingly endless questions. If I have gotten any details wrong, the responsibility is mine.

My literary agent, Laura Wood, helped me develop the story, and provided invaluable guidance during the book proposal process. Thanks also to Stuart Horwitz for his early thoughts, and to Lisa Tener.

One point about literary ethics, wine, and my quest: I paid for all my own travel, food, and lodging. But many, many, people suggested wineries to visit or scientists to talk to, with special thanks to Beth von Benz, Sophia Perpera, and Matti Friedman. Karen MacNeil read a version of the manuscript, and provided valuable feedback.

Luckily the wine gods sent Amy Gash of Algonquin Books to be my editor. She was patient, brilliant, and endlessly supportive. I wouldn't have survived without her. Anne Winslow and Brunson Hoole contributed to design and production and Michael McKenzie and Jacquelynn Burke to marketing and publicity efforts. Michael Newhouse turned my crude sketches into a beautiful map.

Other writers generously read drafts of early chapters, and some of the most valuable feedback came at the Santa Fe Science Writing Workshop. Thanks to instructors Sandra Blakeslee, George Johnson, David Corcoran, and Michael Specter, and to fellow writers Ashwin Budden, Joan Conrow, Sarah Hillenbrand, Mary Dill, Mengxi Zhang, Prashant Nair, and Mariah Salisbury.

Santa Fe delivered an extra bonus, too. I met Cat Warren there. We became friends while talking about North Carolina, science, and writing, and over time she has become an invaluable friend and critic: on my side but tough enough to deliver needed criticisms.

Speaking of North Carolina, many, many years ago Joseph Mitchell, the late *New Yorker* writer, befriended me and encouraged me to try journalism. His writing and his kindness are still an inspiration.

Old friends David Yuro and Terri Hardin helped soothe my doubts. Thanks also to Barbara, Clark, Harry, and Polly Holmes; Deborah Blum; Nancy Albritton; Dawn Sinclair Shapiro; and Anita Gregory. But of course the most devoted reader was my mother, Jane, who still has sharp eyes and a love of books at the age of eighty-six. My father didn't live to see this project, but I remember his stacks of the *New York Review of Books*, and know he would be proud.

Kevin Begos
Apalachicola, Florida

WHERE TO BUY AND LEARN MORE

SOME OF THE wines I wrote about aren't that easy to find. Here are a few importers and retailers who handle a wide variety of vintages, and some knowledgeable critics with fun blogs and newsletters.

Retailers

Kermit Lynch Wine Merchant

1605 San Pablo Ave.
Berkeley, CA 94702
(510) 524-1524

www.kermitlynch.com

Lynch has been importing both classic and lesser-known French and Italian wines since the early 1970s. His online store and newsletters are a great way to learn about different grape varieties.

Astor Wines & Spirits

De Vinne Press Building
399 Lafayette St. (at East Fourth St.)
New York, NY 10003
(212) 674-7500

www.astorwines.com

Astor hosts regular tastings, and you can use their website to search for wines made with hundreds of grapes, including Agiorgitko

(Greece), Boğazkere (Turkey), Picpoul (France), and Zweigelt (Austria).

Chambers Street Wines

148 Chambers St.

New York, NY 10007

212-227-1434

www.chambersstwines.com

A terrific selection of wines from small producers around the world.

Blogs, websites, and other sources of information

José Vouillamoz

@JoseGrapes on Twitter

A never-ending stream of wine grape information and science, filled with posts such as, "My first ever Morokanella, old native #wine #grape from #cyprus in #nicosia," and retweets such as "Get your running shoes on—We've organized the first ever Vineyard trail run in Armenia!!!"

Biomolecular Archaeology Project

Patrick McGovern, University of Pennsylvania

www.penn.museum/sites/biomoleculararchaeology/

Want to know more about "A biomolecular archaeological approach to 'Nordic grog'"? Of course you do. That paper, and details of his teaching, research, and re-creations of ancient ales, are on this hugely fun website.

The Feiring Line

www.alicefeiring.com

A free online blog and subscription newsletter by writer and critic Alice Feiring, full of recommendations, interviews, and insights. Feiring describes her work as "hunting the Leon Trotskys, the Philip Roths, the Chaucers and the Edith Whartons of the wine world. I want them natural and most of all, I want them to speak the truth even if we argue." Feiring is a great source for news on Georgian wines, natural wines, and delightful vintages from all over the world.

Jancis Robinson

www.jancisrobinson.com

Many (including me) consider Robinson to be the most knowledgeable wine critic in the world. She writes beautifully, is incessantly curious, and her website has a tremendous amount of wine information, from Burgundy to China. She has a free newsletter, and reviews by subscription. A number of other talented writers also contribute to the site.

Wine Anorak

www.wineanorak.com

Jamie Goode has a PhD in plant biology, but he left a science job years ago to write about wine. His website has news, features, reviews, and a whole section on wine science. He says "don't be intimidated by some of the seriously geeky bits," and I agree. This is good stuff. (By the way, in this case "anorak" doesn't refer to a polar exploration jacket—it's also British slang for someone who has an obsessive interest in odd subjects.)

RAW WINE

www.rawwine.com

The world's leading celebration of natural wine, held annually in London, has expanded to cities such as New York, Berlin, and Los Angeles. Check their Facebook page for up-to-date news about events: www.facebook.com/rawwineworld.

The Academic Wino

www.academicwino.com

A wonderful blog by scientist and wine enthusiast Becca Yeamans-Irwin on current enology and viticulture research.

The Wineoscope

www.wineoscope.com

The subtitle of this website by yeast researcher Erika Scymanski pretty much sums it up: Wine Science, Wine Rhetoric, and Other Geekery. Remember, we may work for the yeast, not the other way around . . .

CHAPTER NOTES

Chapter 1: A Mysterious Wine

6   **I looked for Cremisan**   *See* Jancis Robinson, ed., *The Oxford Companion to Wine, Third Edition* (New York: Oxford University Press, 2006).

6   **Abu Nuwas, who lived in Baghdad**   *See* Roger M.A. Allen, Encyclopedia Britannica online, *s.v.* "Arabic Literature" (n.d.): https://www.britannica.com/art/Arabic-literature.

7   **From time to time, though**   *See* Peter Washington, ed., *Persian Poets* (New York: Everyman's Library, Alfred A. Knopf, 2000).

7   **Medieval Jewish wine poets**   *See* Raymond Scheindlin, *Wine, Women, and Death: Medieval Hebrew Poems on the Good Life* (New York: Oxford University Press, 1999).

8   **Until I heard about José Vouillamoz's work**   *See* Jancis Robinson, Julia Harding, and José Vouillamoz, *Wine Grapes: A Complete Guide to 1,368 Vine Varieties, Including Their Origins and Flavours* (New York: Ecco Press, 2012).

10   **Thomas Jefferson loved Esopus Spitzenburg apples**   *See* "Esopus SpitzenBurg Apple," Thomas Jeffferson Foundation, https://www.monticello.org/site/house-and-gardens/in-bloom/esopus-spitzenburg-apple.

10   **One called Cavendish occupies 90 percent**   *See* Roberto A. Ferdman, "Bye, bye bananas," *Washington Post*, Dec. 4, 2015.

10   **The *Atlantic* published the article**   *See* Robert M. Parker Jr., "The Dark Side of Wine," *Atlantic*, Dec. 2000, originally published in *Parker's Wine Buyer's Guide, Fifth Edition* (New York: Simon & Schuster, 1999).

10   **Statistics show there is cause to worry**   *See* Kym Anderson, "Changing Varietal Distinctiveness of the World's Wine Regions: Evidence from a New Global Database," *Journal of Wine Economics* 9, no. 3 (Nov. 3, 2014): 249–72.

12   **One day I visited the *Wine Grapes* website**   *See* Jancis Robinson, Julia Harding, and José Vouillamoz, "#NotInWineGrapes competition," Wine Grapes, Feb. 2013, http://winegrapes.org/notinwinegrapes-contest/.

Chapter 2: Archaeobiology and Ancient Wine

14   **"[A mass spectrometer] weighs molecules"**   *See* "Critical Mass: A History of Mass Spectrometry," Chemical Heritage Foundation, http://archive.is/3O5mZ; *see also* Scripps Center for Metabolomics, "Basics of Mass Spectrometry," Scripps Research Institute website, https://masspec.scripps.edu/landing_page.php?pgcontent=whatIsMassSpec.

14   **During a lecture in the 1960s**   *See The Feynman Lectures on Physics; The Complete Audio Collection, Vol. 11* (New York: Basic Books, 2007).

15   **Processes such as liquid chromatography**   *See* NMSU College of Arts and Sciences, Department of Chemistry and Biochemistry, "Liquid Chromatography," New Mexico State University, https://web.nmsu.edu/~kburke/Instrumentation/Lqd_Chroma.html.

15   **Spanish researchers identified three types**   *See* Irep en Kemet Project: Documenting the Corpus of Wine in Ancient Egypt website, Barcelona: Archaeological Institute of America, https://www.archaeological.org/fieldwork/cp/10941.

15   **One of McGovern's projects looked at residue**   *See* Patrick McGovern, Armen Mirzoian, and Gretchen R. Hall, "Ancient Egyptian herbal wines," *Proceedings of the National Academy of Sciences* 106, no. 18 (May 2009): 7361–6.

17   **To find out where they were manufactured**   *See* Patrick McGovern, *Ancient Wine: The Search for the Origins of Viniculture* (Princeton, NJ: Princeton University Press, 2003).

17   **That research lends credence to a passage . . . Herodotus wrote**   *See* Herodotus, *The History of Herodotus*, trans. George Rawlinson (London: John Murray, 1858).

18   **The Egyptians also wrote poems about wine**   *See* Michael V. Fox, *The Song of Songs and the Ancient Egyptian Love Songs* (Madison: University of Wisconsin Press, 1985).

18   **That poem led me back even further**   *See* Jeremy Black et al., *The Electronic Text Corpus of Sumerian Literature*, Faculty of Oriental Studies, University of Oxford (Nov. 30, 2016): http://etcsl.orinst.ox.ac.uk/.

18   **And this passage is from *The Tale of Sinuhe***   *See* Miriam Lichtheim, *Ancient Egyptian Literature: A Book of Readings, Vol. I, The Old and Middle Kingdoms* (Berkeley: University of California Press, 1973).

19   **My vague notions of ancient wine were expanding**   *See* Patrick McGovern, *Ancient Wine* (Princeton, NJ: Princeton University Press, 2003).

19   **I even smiled at ancient jokes**   *See* Jeremy Black et al., *The Electronic Text Corpus of Sumerian Literature*, Faculty of Oriental Studies, University of Oxford (Nov. 30, 2016): http://etcsl.orinst.ox.ac.uk/.

19 **Other researchers used a somewhat bizarre contraption**   *See* Ulrike Heberlein et al., "Molecular Genetic Analysis of Ethanol Intoxication in *Drosophila melanogaster*," *Integrative & Comparative Biology* 44, no. 4 (Aug. 1, 2004): 269–74.

20 **Elephants raiding Indian breweries have inspired headlines**   *See* Eli MacKinnon, "Is Every Single Elephant a Village-Wrecking Booze Hound?", Live Science website, Nov. 9, 2012, http://www.livescience .com/24678-is-every-single-elephant-a-village-wrecking-booze-hound .html.

21 **In 1999 McGovern led a team that used archaeobiology**   *See* Patrick McGovern et al., "A funerary feast fit for King Midas," *Nature* 402 (Dec. 23, 1999): 863–4.

23 **Two passages in the Babylonian Code of Hammurabi**   *See* L. W. King, trans., *The Code of Hammurabi*, The Avalon Project, Yale Law School, http://avalon.law.yale.edu/ancient/hamframe.asp.

23 **Around 360 BC Plato described various**   *See* Plato, *Laws*, trans. Benjamin Jowett (Oxford University: Clarendon Press, 1875).

24 **Throughout the third to fifth centuries AD**   *See* Michael L. Rodkinson, trans., *New Edition of the Babylonian Talmud, Volume III: Section Moed (Festivals): Tract Erubin* (Boston: New Talmud Publishing Company, 1903).

24 **Ancient clay tablets from Mesopotamia**   *See* Wolfgang Heimpel, *Letters to the King of Mari: A New Translation, with Historical Introduction, Notes, and Commentary* (Winona Lake, Indiana: Eisenbrauns, 2003); *see also* A. Leo Oppenheim, trans., *Letters from Mesopotamia: Official, Business, and Private Letters on Clay Tablets from Two Millennia* (Chicago: University of Chicago Press, 1967).

25 **Wine was important enough that the Hittite army**   *See* John Campbell, *The Hittites: Their Inscriptions and Their History, Vol. II* (London: John C. Nimmo, 1891).

Chapter 3: Cremisan

28 **He is a renowned archaeologist**   *See* Maier, Aren, Tell es-Safi/Gath Archaeological Project Official (and Unofficial) Weblog, https://gath .wordpress.com/.

29 **Later I casually looked up Emek Refaim Street**   *See* Philologos, "Ghostly," *Forward*, March 26, 2004.

38 **The venerable Jewish–American newspaper**   *See* Philologos, "The Adorable Moses Cow," *Forward*, April 23, 2004.

Chapter 4: Israel's Forgotten Grapes

43   **When I first read about Drori's work**   *See* Ariel University Wine Research Center, "Reviving the Wines of Ancient Israel," Friends of Ariel University website, http://www.afau.org/blog/2015/05/14/the-ariel -univeresity-wine-research-center-reviving-the-wines-of-ancient-israel/.

45   **In 1997 German researchers did the same thing**   *See* Manfred Heun et al., "Site of Einkorn Wheat Domestication Identified by DNA Finger-printing," *Science* 278, iss. 5341 (Nov. 14, 1997): 1312–4.

47   **Sean Myles, the lead author on a widely cited 2011 paper**   *See* Sean Myles et al., "Genetic structure and domestication history of the grape," *Proceedings of the National Academy of Sciences* 108, no. 9 (Jan. 18, 2011): 3530–5.

48   **As UC Davis scientist Carole Meredith observed**   *See* Esther Mobley, "New wine film 'Somm: Into the Bottle' is ambitious, dangerously selective," SFGate website, Nov. 12, 2015, http://insidescoopsf.sfgate.com /blog/2015/11/12/new-wine-film-somm-into-the-bottle-is-ambitious -dangerously-selective/.

49   **Whether you're writing about finches, tortoises, earthworms**   *See* Charles Darwin, *The Movements and Habits of Climbing Plants* (London: John Murray, 1875).

51   **The earliest lianas survived the K–T mass extinction**   *See* Norman Sleep and Donald Lowe, "Scientists reconstruct ancient impact that dwarfs dinosaur-extinction blast," American Geophysical Union website, Washington, DC, April 9, 2014, https://news.agu.org/press-release /scientists-reconstruct-ancient-impact-that-dwarfs-dinosaur-extinction -blast/.

Chapter 5: The Wine Scientist

56   ***Decanter* called it**   *See* Andrew Jefford, "Jefford on Monday: The Grammar of Wine," *Decanter* website, Oct. 22, 2012, http://www.decanter.com /wine-news/opinion/jefford-on-monday/jefford-on-monday-the-grammar -of-wine-24418/; see also Eric Asimov, "Highlights From the 2012 Vintage in Wine Publishing," *New York Times*, Nov. 30, 2012.

59   **It is in an Armenian cave**   *See* Hans Barnard et al., "Chemical evidence for wine production around 4000 BCE in the Late Chalcolithic Near Eastern highlands," *Journal of Archaeological Science* 38, iss. 5 (May 2011): 977–84.

60   **In 2013 French researchers compared DNA samples**   *See* Roberto Bacilieri et al., "Genetic structure in cultivated grapevines is linked to geography and human selection," *BMC Plant Biology* 13, no. 25 (Feb. 8, 2013).

60 **Ehud Weiss, an expert on plant genetics**   *See* Ehud Weiss, "'Beginnings of Fruit Growing in the Old World' two generations later," *Israel Journal of Plant Sciences* 62, nos. 1–2 (2015): 75–85.

62 **That means we essentially have no way of knowing**   *See* Pliny the Elder, *The Natural History of Pliny, Vol. III, Ch. 20*, trans. John Bostock and H. T. Riley (London: Henry G. Bohn, 1855).

63 **The attraction still exists**   *See* Anna Schneider, et al., *Ampelografia Universale Storica Illustrata: I Vitigni del Mondo* (Turin: L'Artistica Editrice, 2012).

66 **Meredith put the birth in perspective**   *See* Carole Meredith, interview by Geoff Kruth and Matt Stamp, "An Interview with Dr. Carole Meredith," GuildSomm website, November 2011, https://www.guildsomm.com/stay_current/features/b/guest_blog/posts/the-science-of-grape-genetics-an-interview-with-dr-carole-meredith.

66 **When the news about Chardonnay's parents broke**   *See* Nicholas Wade, "For a Noble Grape, Disdained Parentage," *New York Times*, Sept. 3, 1999.

66 **A battle over Zinfandel's origins**   *See* Lynn Alley, "ATF Proposes Primitivo as Synonym for Zinfandel," *Wine Spectator*, April 16, 2002.

67 **The global market is worth**   *See* "2016 World wine production estimated at 259 mhl," OIV life website, Oct. 20, 2016, http://www.oiv.int/en/oiv-life/2016-world-wine-production-estimated-at-259-mhl.

68 **We hear about DNA all the time**   *See* "About the MBL Logan Science Journalism Program," Marine Biological Laboratory website, University of Chicago, http://www.mbl.edu/sjp/.

71 **Much later, subsidies from the European Union**   *See* Lisa Hirai, "Distillation: An Effective Response to the Wine Surplus in the European Community?", *Boston College International and Comparative Law Review* 16, iss. 1 (Dec. 1, 1993); *see also* Barrett Ludy, "ConfEUsion: A Quick Summary of the EU Wine Reforms," GuildSomm website, Oct. 5, 2012.

72 **In 2006 he discovered that Sangiovese**   *See* J. F. Vouillamoz et al., "The parentage of 'Sangiovese', the most important Italian wine grape," *Vitis* 46, no. 1 (2007): 19–22.

75 **I was at VinEsch, near the town of Visp**   *See* VinEsch website (auto-translated from German), http://www.vinesch.ch.

79 **"Corked," he said, using the term for a bottle**   *See* Maria Luisa Álvarez-Rodríguez et al., "Cork Taint of Wines: Role of the Filamentous Fungi Isolated from Cork in the Formation of 2,4,6-Trichloroanisole by O Methylation of 2,4,6-Trichlorophenol," *Applied and Environmental Microbiology* 68, no. 12 (Dec. 2002): 5860–9.

Chapter 6: Flavor, Taste, and Money

84　**Shepherd and others use real-time brain scans**　*See* Gordon Shepherd, "Neuroenology: how the brain creates the taste of wine," *Flavour* 4, no. 19 (March 2, 2015).

85　**In 2014 French researchers monitored**　*See* Lionel Pazart et al., "An fMRI study on the influence of sommeliers' expertise on the integration of flavor," *Frontiers in Behavioral Neuroscience* 8, no. 358 (Oct. 2014).

85　**In one field study a display of French and German wines**　*See* A. C. North, David J. Hargreaves, and Jennifer McKendrick, "The Influence of In-Store Music on Wine Selections," *Journal of Applied Psychology* 84, no. 2 (April 1999): 271–6.

86　**In another study twenty-six people tasted three different wines**　*See* Charles Spence et al., "Looking for crossmodal correspondences between classical music and fine wine," *Flavour* 2, no. 29 (Dec. 19, 2013).

86　**But what about the grapes themselves?**　*See* Olivier Jaillon et al., "The grapevine genome sequence suggests ancestral hexaploidization in major angiosperm phyla.", *Nature* 449 (Sept. 27, 2007): 462–7.

86　**Laboratories can now detect**　*See* Claudia Wood et al., "From Wine to Pepper: Rotundone, an Obscure Sesquiterpene, Is a Potent Spicy Aroma Compound," *Journal of Agricultural and Food Chemistry* 56, no. 10 (ACS Publications, May 8, 2008): 3738–44.

87　**Jamie Goode, the British wine writer, tells of one professor**　*See* Jamie Goode, *The Science of Wine: From Vine to Glass, Second Edition* (Berkeley: University of California Press, 2014).

87　**A Stanford University study offered the same wine**　*See* Hilke Plassmann et al., "Marketing actions can modulate neural representations of experienced pleasantness," *Proceedings of the National Academy of Sciences* 105, no. 3 (Jan. 22, 2008): 1050–4.

88　**In another experiment**　*See* Brian Wansink, Collin R. Payne, and Jill North, "Fine as North Dakota Wine: Sensory Expectations and the Intake of Companion Foods," *Physiology & Behavior* 90, iss. 5 (April 23, 2007): 712–6.

90　**Pliny the Elder wrote this**　*See* Pliny the Elder, *The Natural History of Pliny, Vol. III, Book XIV*, trans. John Bostock and H. T. Riley (London: Henry G. Bohn, 1855).

90　**"I shall not attempt, then, to speak of every kind of vine**　*See* Pliny the Elder, *The Natural History of Pliny, Vol. III*, trans. John Bostock and H. T. Riley (London: Henry G. Bohn, 1855).

90　**The Greek writer Athenaeus**　*See* Athenaeus, *The Deipnosophists*, trans. Charles Burton Gulick (Cambridge, MA: Harvard University Press, 1927).

91 **Maeir said recent excavations in Israel point** *See* Andrew J. Koh, Assaf Yasur-Landau, and Eric H. Cline, "Characterizing a Middle Bronze Palatial Wine Cellar from Tel Kabri, Israel," *PLOS ONE* (Public Library of Science, Aug. 27, 2014).

91 **Wine "labels" turned out to be far older** *See* Eva-Lena Wahlberg, *The Wine Jars Speak: A text study* (Uppsala, Sweden: Uppsala University Department of Archaeology and Ancient History, 2012).

Chapter 7: The Caucasus

95 **Though Georgia is mostly politically stable** *See* "Country information/ Russia," US Department of State website, accessed Jan. 12, 2017, https:// travel.state.gov/content/passports/en/country/russia.html (warning has since been removed).

95 **Anthropologist Florian Mühlfried spent time with them** *See* Florian Mühlfried, *Being a State and States of Being in Highland Georgia* (New York: Berghahn Books, 2014).

98 **The festival is a touchy subject for the Alaverdi monks** *See* Giorgi Shengelaia, dir., *Alaverdoba* (1962): http://www.geocinema.ge/en/index .php?filmi=173.

100 **In the 1920s Nikolai Vavilov** *See* Encyclopedia Britannica online, *s.v.* "Nikolay Ivanovich Vavilov, Russian geneticist" (1998): https://www .britannica.com/biography/Nikolay-Ivanovich-Vavilov.

101 **An expedition from the Chicago Botanic Garden** *See* "Plant Expedi- tion to the Republic of Georgia—Caucasus Mountains," Chicago Botanic Garden website, 2010, https://www.chicagobotanic.org/downloads /collections/georgia2010.pdf.

101 **The region came into existence** *See* Lewis Owen, Nikolay Andreyevich Gvozdetsky, and Solomon Ilich Bruk, Encyclopedia Britannica online, *s.v.* "Caucasus, Region And Mountains, Eurasia" (2015): https://www .britannica.com/place/Caucasus.

101 **As a United Nations report on genetic diversity notes** *See* Caterina Batello et al., "At the Crossroads Between East and West" in *Gardens of Biodiversity: Conservation of genetic resources and their use in traditional food production systems by small farmers of the Southern Caucasus*, ed. Roberta Mitchell (Rome: Food and Agriculture Organization of the United Nations, 2010).

104 **Some writers use the term "orange wine"** *See* Carson Demmond, "Forget Red, White, and Rosé—Orange Wine Is What You Should Be Sipping This Fall," *Vogue*, Oct. 5, 2015.

104 **Levi Dalton, who has worked with legendary chefs** *See* Anna Lee C. Iijima, "Why Orange Wines Will Never Be Mainstream—But a case for why they're more than a dying trend," *Wine Enthusiast*, March 5, 2013.

104 **The British wine writer Simon Woolf**   *See* Simon J. Woolf, "2010 Ala-verdi Monastery Rkatsiteli, Kakheti," Tim's Tasting Notes website, Jan. 1, 2013, http://www.timatkin.com/reviews?807.

106 **The critic and writer Alice Feiring tells of a German scientist**   *See* Alice Feiring, "Random thoughts on Georgia," RAW WINE website, Apr. 13, 2015, http://www.rawwine.com/blog/random-thoughts-georgia.

107 **Now Honnef is the winemaker**   *See* Château Mukhrani website, http://chateaumukhrani.com/en/home.

111 **The most notorious example**   *See* Joel Mokyr, Encyclopedia Britannica online, *s.v.* "Great Famine, Famine, Ireland [1845–1849]" (April 19, 2017): https://www.britannica.com/event/Great-Famine-Irish-history.

111 **The Greek historian Herodotus described**   *See* Herodotus, *The History of Herodotus*, trans. George Rawlinson (London: John Murray, 1858).

112 **An obscure book added even more color**   *See* John Colarusso, trans., *Nart Sagas: Ancient Myths and Legends of the Circassians and Abkhazians* (Princeton, NJ: Princeton University Press, 2016).

Chapter 8: Yeast, Co-Evolution, and Wasps

116 **Master of Wine Sally Easton took part**   *See* Sally Easton, "Saccharomyces interspecific hybrids: a new tool for sparkling winemaking," WineWisdom website, June 20, 2016, http://www.iccws2016.com/wp-content/uploads/2013/07/Wine-Wisdom-206.pdf.

117 **Spanish researchers found**   *See* J. E. Perez-Ortin et al., "Molecular Characterization of a Chromosomal Rearrangement Involved in the Adaptive Evolution of Yeast Strains," *Genome Research* 12, no. 10 (October 2002): 1533–9.

118 **Yeast use our global wine-lust to travel, too**   *See* M. R. Goddard et al., "A distinct population of Saccharomyces cerevisiae in New Zealand: evidence for local dispersal by insects and human-aided global dispersal in oak barrels.", *Environmental Microbiology* 12, no. 1 (January 2010): 63–73.

118 **Some crafty *S. cerevisiae* found a solution**   *See* Irene Stefanini et al., "Role of social wasps in *Saccharomyces cerevisiae* ecology and evolution," *Proceedings of the National Academy of Sciences* 109, no. 33 (Aug. 14, 2012): 13398–403.

119 **French researchers examined yeast DNA**   *See* Delphine Sicard and J. L. Legras, "Bread, beer and wine: yeast domestication in the Saccharomyces sensu stricto complex.", *Comptes Rendus Biologies* 334, no. 3 (March 2011): 229–36.

119 **Another team of researchers looked at modern yeast DNA**   *See* Krista Conger, "In vino veritas: Promiscuous yeast hook up in wine-making vats, study shows," Stanford Medicine News Center, Feb. 26, 2012.

120   **There's still disagreement over where yeasts originated**   *See* Raúl J. Cano et al., "Amplification and sequencing of DNA from a 120–135-million-year-old weevil," *Nature* 363 (June 10, 1993): 536–8.

120   **But change did come, in 1965**   *See* Erika Szymanski, "Spontaneous Fermentation: Bubble, Bubble, Less Toil, or Trouble?" Palate Press website, July 25, 2010, http://palatepress.com/2010/07/wine /spontaneous-fermentation-wine-bubble-bubble-less-toil-or-trouble/.

121   **Commercial wine yeasts are taking over**   *See* Gemma Beltran et al., "Analysis of yeast populations during alcoholic fermentation: A six year follow-up study," *Systematic and Applied Microbiology* 25, iss. 2 (2002): 287–93.

121   **A quirky twist left me hopeful and puzzled**   *See* Gabe Oppenheim, "The Beer That Takes You Back . . . Millions of Years," *Washington Post*, Sept. 1, 2008.

121   **In 2008 Cano started offering trial batches**   *See* Ian Schuster, "45-million-year-old-yeast beer Early Access," Indiegogo, https://www .indiegogo.com/projects/45-million-year-old-yeast-beer-early-access#/.

121   **Schubros brewer Ian Schuster told**   *See* Alyssa Pereira, "The East Bay beer that's 45 million years old," SFGate, Aug. 26, 2016, http://www .sfgate.com/food/article/The-East-Bay-beer-that-s-45-million-years -old-9177673.php.

122   **Clayton Cone, a scientist with Lallemand**   *See* Clayton Cone, "Yeast Genetics and Flavor," Lallemand website, March 25, 2017, http://beer .lallemandyeast.com/articles/yeast-genetics-and-flavor.

122   **The US Food and Drug Administration approved**   *See* Prof. Joe Cummins, "Genetically Engineered Wine & Yeasts Now on the Market," Organic Consumers Association website, Dec. 1, 2005, https://www .organicconsumers.org/old_articles/ge/wine121005.php.

123   **Szymanski wrote in a blog post**   *See* Erika Szymanski, "Has Yeast Domesticated Us?", Palate Press: The Online Wine Magazine, March 10, 2013, http://palatepress.com/2013/03/wine/has-yeast-domesticated-us/.

Chapter 9: Aphrodite, Women, and Wine

124   **She is . . . the moon and . . . men sacrifice**   *See* Macrobius, *Saturnalia: Books 3–5*, trans. Robert A. Kaster (Cambridge, MA: Loeb Classical Library, Harvard University Press, 2011).

125   **The earliest nymphaeum**   *See* "Nymphaeum, Ancient Greco-Roman Sanctuary," Encyclopedia Britannica online (2008): https://www .britannica.com/art/nymphaeum.

125   **The pharaohs of ancient Egypt came there**   *See* Eric H. Cline, ed., *The Oxford Handbook of the Bronze Age Aegean (ca. 3000–1000 BC)* (New York: Oxford University Press, 2010).

127 **In the medieval French poem**   *See* Ben O'Donnell, "The First Wine Competition?", *Wine Spectator,* May 31, 2011.

135 **Patrick McGovern found that both human and divine females**   *See* Patrick McGovern, *Uncorking the Past: The Quest for Wine, Beer, and Other Alcoholic Beverages* (Berkeley: University of California Press, 2010); *see also* Patrick McGovern, *Ancient Wine* (Princeton, NJ: Princeton University Press, 2003).

136 **Siduri is a major character in** ***The Epic of Gilgamesh***   *See* Andrew George, trans., *The Epic of Gilgamesh* (New York: Penguin Books, 2003).

136 **"[It] seems evident that participation**   *See* Ross S. Kraemer, "Ecstasy and Possession: The Attraction of Women to the Cult of Dionysus," *Harvard Theological Review* 72, nos. 1–2 (Jan.–Apr. 1979): 55–80.

138 **Juvenal, a Roman satirist**   *See* G. G. Ramsay, trans., *Juvenal and Persius* (New York: G. P. Putnam's Sons, 1918).

138 **He later complained**   *See* Sarah B. Pomeroy, *Goddesses, Whores, Wives, and Slaves: Women in Classical Antiquity* (New York: Schocken Books, 1995).

Chapter 10: Goliath, Foraging, and an Answer

143 **I arrived in Israel again**   *See* Times of Israel Staff, "Worst air pollution ever in Jerusalem as sandstorm engulfs Mideast," *Times of Israel,* Sept. 8, 2015.

Chapter 11: Italy, Leonardo, and Natural Wine

155 **Giusto was a pioneer, too**   *See* Ivan Brincat, "Azienda Agricola COS: A Sicilian winemaker with a difference," *Food and Wine Gazette,* Dec. 8, 2014.

155 **In 2003 she sent a manifesto letter**   *See* Arianna Occhipinti, *Natural Woman: My Sicily, My Wine, My Passion* (Rome: Fandango Libri, 2013); original in Italian with English translation by author.

156 **I thought of an interview she'd given**   *See* Arianna Occhipinti, interview by Charles Gendrot at Cork and Fork wine event, Washington, D.C., April 9, 2015, http://washingtondc.eventful.com/events/dc -arianna-occhipinti-wines-frappato-nero-davol-/E0-001-081472887-0.

160 **"Wine and table grapes currently receive**   *See* Sean Myles et al., "Genetic structure and domestication history of the grape," *Proceedings of the National Academy of Sciences* 108, no 9 (March 1, 2011): 3530–5.

160 **Isabelle Legeron is often seen as the leader**   *See* Isabelle Legeron MW, "What is Natural Wine?" *Decanter* magazine via RAW WINE website, Sept. 2011, http://www.rawwine.com/what-natural-wine.

160 **Cecilia Díaz, a German environmental and food scientist** *See* Cecilia Díaz et al., "Characterization of Selected Organic and Mineral Components of Qvevri Wines," *American Society for Enology and Viticulture* 64 (Dec. 2013): 532–7.

161 **Roberto Ferrarini, an Italian scientist** *See* Ferrarini, Roberto, "The 'Long Time Skin Contact' Typical Technique of Qvevri Wines: Its Effects on Phenolic and Aromatic Wine Composition," 1st International Qvevri Wine Symposium Report Georgian Wine Association website, Sept. 11, 2011, http://doczz.fr/doc/2201964/1st-international-qvevri -wine-symposium.

161 **The critic Simon Woolf took a deeper look** *See* Simon Woolf, "Georgian qvevri wine: if it's good enough for God . . .", timatkin.com, Jan. 1, 2013, http://www.timatkin.com/articles?803.

162 **In 2014 Parker addressed a California wine conference** *See* David White, "Robert Parker Responds to Jon Bonné," Terroirist: a daily wine blog, Feb. 25, 2014, http://www.terroirist.com/2014/02 /robert-parker-responds-to-jon-bonne/.

163 **I scanned local papers** *See* La Vigna di Leonardo website, http://www .vignadileonardo.com/setlang=en.

164 **A sixteenth-century map of Milan** *See* Charles Nicholl, *Leonardo da Vinci: Flights of the Mind* (New York: Viking, 2004).

164 **In 1507 the French asked him to return** *See* Charles Nicholl, *Leonardo da Vinci: Flights of the Mind*

168 **Whether you believe Jesus** *See* Michael Harthorne, "The Last Supper of Jesus Didn't Happen at a Table," Newser, March 25, 2016.

Chapter 12: Wine and Foie Gras

172 **Archaeological and DNA evidence show** *See* Patrick E. McGovern et al., "Beginning of viniculture in France," *Proceedings of the National Academy of Sciences* 110, no. 25 (June 18, 2013): 10147–52.

173 **A 1310 manuscript by the Franciscan Vital du Four** *See* Relax News, "France vaunts '40 virtues' of Armagnac," *Independent*, Feb. 7, 2010.

174 **My first stop with Caillard was a 190-year-old vineyard** *See* Marcel Michelson, "French classify ancient vines as national treasure," *Reuters*, Paris, June 26, 2012.

185 **That brought to mind Henry James's reaction to Bordeaux** *See* Henry James, *A Little Tour of France* (Cambridge, MA: The Riverside Press, Houghton, Mifflin and Company, 1900).

186 **As of 2016 there were more than two hundred thousand acres** *See* Sylvia Wu, "Cabernet Sauvignon reigns in Chinese wine regions, shows

report," *Decanter*, Dec. 22, 2015; *see also* L. I. Demei, Chinese wines made from native furry vines, *Decanter China*, Feb. 17, 2015

Chapter 13: The Science of Terroir

190   **"The history of *terroir* itself . . .**   *See* Rod Phillips, "The Myths of French Wine History," GuildSomm website, Oct. 17, 2016, https://www.guildsomm.com/stay_current/features/b/rod_phillips/posts/french-wine-myths.

191   **Alex Maltman, a geologist**   *See* Alex Maltman, "The Role of Vineyard Geology in Wine Typicity," *Journal of Wine Research* 19, iss. 1 (Aug. 13, 2008): 1–17.

191   **There is a long history of wine lovers**   *See* R. S. "Food at the Exposition," *New York Times*, Aug. 12, 1900.

191   **A 1988 *Times* article about a wine conference**   *See* Howard G. Goldberg, "Bordeaux Team Comes to L.I. To Share Expertise," *New York Times*, August 3, 1988.

191   **Then a 1993 *New York Times* wine review**   *See* "Tastings," *New York Times*, Feb. 24, 1993.

192   **A 2014 study of vineyards all over California**   *See* Nicholas A. Bokulich et al., "Microbial biogeography of wine grapes is conditioned by cultivar, vintage, and climate," *Proceedings of the National Academy of Sciences* 111, no. 1 (Jan. 7, 2014): 139–48.

193   **The *SOMM Journal* remarked**   *See* David Gadd, "The Reading Room," *Somm Journal*, Aug.–Sept. 2016: 11.

Chapter 14: Coming Home, and Holy Land Wine

195   **Granick is a former Fulbright scholar**   *See* The Institute of the Masters of Wine website, *s.v.* "Lisa Granik," http://www.mastersofwine.org/en/meet-the-masters/profile/index.cfm/id/6fb5a043-5e4b-e211-a20600155d6d822c.

196   **Around the same time I wrote a travel story**   *See* Kevin Begos / Associated Press, "Palestinian winemakers preserve ancient traditions," *San Diego Union-Tribune*, Oct. 13, 2015.

196   **That November the *New York Times* profiled an Israeli wine**   *See* Jodi Rudoren, "Israel Aims to Recreate Wine That Jesus and King David Drank," *New York Times*, Nov. 29, 2015.

196   **In December CNN.com featured both the Israeli and Palestinian wines**   *See* Oren Liebermann, "What would Jesus drink?", CNN website, Dec. 23, 2015, http://www.cnn.com/2015/12/23/middleeast/jesus-wine/.

197 **It was time to read and reflect** *See* Rod Phillips, "The Myths of French Wine History," GuildSomm website, Oct. 17, 2016, https://www.guildsomm.com/stay_current/features/b/rod_phillips/posts/french-wine-myths.

198 **In the early 1880s** *See* Simon Schma, *Two Rothschilds and the Land of Israel*, (New York: Alfred A. Knopf, 1978.)

199 **In 1453 Sultan Mehmet II openly courted Jewish immigrants** *See* Rachel Avraham, "Ottoman Empire: A Safe Haven for Jewish Refugees," JerusalemOnline, June 11, 2014, http://www.jerusalemonline.com/israel-history/ottoman-empire-a-safe-haven-for-jewish-refugees-5797.

200 **Historian Avigdor Levy found that Nasi** *See* Avigdor Levy, ed., *Jews, Turks, Ottomans: A Shared History, Fifteenth Through the Twentieth Century* (Syracuse, NY: Syracuse University Press, 2002).

200 **I found records from Cyprus showing** *See* Michalis N. Michael, Matthias Kappler, and Eftihios Gavriel, eds., *Ottoman Cyprus: A Collection of Studies on History and Culture* (Wiesbaden, Germany: Harrassowitz Verlag, 2009).

Chapter 15: American Wine Grapes

204 **Dufour emigrated from Switzerland** *See* Thomas Pinney, "Dufour and the Beginning of Commercial Production" in *A History of Wine in America: From the Beginnings to Prohibition* (Berkeley: University of California Press, 1989): 117–26.

208 **"A few years ago, I never imagined** *See* Eric Asimov, "A Top 10 Wine List So Good, It Takes 12 Bottles to Hold It," *New York Times*, Dec. 10, 2015.

209 **"[W]e knew it would be an enormous undertaking** *See* Eric Asimov, "At La Garagista, Hybrid Grapes Stand Up to Vermont's Elements," *New York Times*, Aug. 27, 2015.

212 **Yet Grahm talks openly about his failures** *See* Eric Asimov, "His Big Idea Is to Get Small," *New York Times*, April 21, 2009.

213 **In early 2016 Grahm summed it all up** *See* Bill Zacharkiw, "California dreaming with Bonny Doon winemaker Randall Grahm," *Montreal Gazette*, March 24, 2016.

215 **A leading California newspaper used the term "Frankengrapes"** *See* Esther Mobley, "Can Andy Walker save California wine?", *San Francisco Chronicle*, Jan. 29, 2016.

217 **Grahm summed it up at a Brooklyn food conference** *See* Randall Grahm, "Speech presented at the Food + Enterprise Conference," Bonny Doon Vineyard website, May 5, 2015, https://www.bonnydoonvineyard.com/food-enterprise-speech/.

219   **Berry, the Kentucky poet**   *See* Wendell Berry, "Confessions of a Water Drinker" in *Inspiring Thirst: Vintage Selections from the Kermit Lynch Wine Brochure*, edited by Kermit Lynch (Berkeley, CA: Ten Speed Press, 2004).

222   *Forbes* **published an article online**   *See* Cathy Huyghe, "Coming Soon To A Winery Near You: Ancient Amphoras," *Forbes* website, Feb. 24, 2014, https://www.forbes.com/sites/cathyhuyghe/2014/02/24/coming-soon-to-a-wine-near-you-ancient-amphoras/#6da3fb9d2491; *see also* Chelsea Morse, "An Artist-Turned-Winemaker's Incredible Amphoras," *Food & Wine* website, http://www.foodandwine.com/blogs/artist-turned-winemakers-incredible-amphoras; *see also* Kerin O'Keefe, "Ancient Vessels, Modern Wines," *Wine Enthusiast* website, Aug. 3, 2016.

222   **An Oregon wine critic wrote**   *See* Katherine Cole, "The ancient art of terracotta-fermented wines gets new life in Oregon: Wine Notes," *Oregonian* via OregonLive.com, Aug. 13, 2014, http://www.oregonlive.com/foodday/index.ssf/2014/08/ancient_art_makes_for_antiquat.html.

225   **Here's how he described**   *See* "2015 SM3, Minimus," Craft Wine Co. website, https://craft-wine-co.myshopify.com/products/2015-sm3.

Chapter 16: The Dark Side of Wine Science

231   **Take Ava Winery**   *See* www.avawinery.com.

232   *New Scientist* **compared it to Italian Ruffino**   *See* Chris Baraniuk, "Synthetic wine made without grapes claims to mimic fine vintages," *New Scientist*, May 16, 2016.

233   **Around the same time Gïk released its wine**   *See* Julie Bensman, "The world's first blue wine," BBC Travel, Nov. 19, 2016.

233   **Another new product announcement floored me**   *See* www.vinome.com.

234   **The Helix press release about Vinome opened**   *See* "Helix Announces New Partnerships with National Geographic and Mount Sinai," Business Wire, Oct. 26, 2016.

234   **The health and medicine news agency STAT asked geneticist**   *See* Rebecca Robbins, "This startup claims to pair different wines with your DNA," *Business Insider*, Oct. 27, 2016.

235   **Steven Shapin, a History of Science professor**   *See* "The Tastes of Wine: Towards a Cultural History," *wineworld: new essays on wine, taste, philosophy and aesthetics* 51, anno LII, March 2012.

235   **Today the US Department of the Treasury operates**   *See* "Beverage Alcohol Laboratory" US Department of the Treasury Alcohol and Tobacco Tax and Trade Bureau website, https://www.ttb.gov/ssd/beverage_alcohol_lab.shtml.

Chapter 17: Grapetionary

238 **Tesauro created an event called Grapetionary A–Z** *See* grapetionary: vines and wines A–Z website, http://www.grapetionary.com/.

240 **A *Forbes* reviewer called it** *See* Huyghe, Cathy, "The Most Unique Wine Event I've Ever Experienced," *Forbes* website, June 9, 2016, https://www.forbes.com/sites/cathyhuyghe/2016/06/09/the-most-unique-wine-event-ive-ever-experienced/#65610c496fc5.

240 **A *Washington Post* headline read** *See* Dave McIntyre, "From aglianico to zibibbo, he poured an alphabetical tour of wine grapes," *Washington Post*, Oct. 29, 2016.

Chapter 18: The First Grapes, and Tasting the Past

245 **The museum visit reminded me** *See* Jane Kramer, "A Feast for 'Jerusalem' at the Met," *New Yorker*, Dec. 4, 2016.

248 **Manchester's team eventually found** *See* Steven R. Manchester, Dashrath K. Kapgate, and Jun Wen, "Oldest fruits of the grape family (Vitaceae) from the Late Cretaceous Deccan cherts of India.", *American Journal of Botany* 100, no. 9 (Sept. 2013): 1849–59.

250 **Drori's research creates a foundation** *See* Elyashiv Drori et al., "Collection and characterization of grapevine genetic resources (*Vitis vinifera*) in the Holy Land, towards the renewal of ancient winemaking practices," *Scientific Reports* 7, no. 44463 (March 17, 2017).

251 **Adrian Newton, a scientist who has studied the apples** *See* A. Eastwood et al., *The Red List of Trees of Central Asia*, Fauna & Flora International, 2009; *see also* Josie Glausiusz, "Apples of Eden: Saving the Wild Ancestor of Modern Apples," *National Geographic*, May 9, 2014.

BIBLIOGRAPHY

Campbell, Christy. *The Botanist and the Vintner: How Wine Was Saved for the World*. Chapel Hill, NC: Algonquin Books of Chapel Hill, 2006.

Colarusso, John, translator. *Nart Sagas from the Caucasus: Myths and Legends from the Circassians, Abazas, Abkhaz, and Ubykhs*. Princeton, NJ: Princeton University Press, 2002.

Cole, Katherine. *Voodoo Vintners: Oregon's Astonishing Biodynamic Winegrowers*. Corvallis: Oregon State University Press, 2011.

Feiring, Alice. *For the Love of Wine: My Odyssey Through the World's Most Ancient Wine Culture*. Lincoln, NE: Potomac Books, 2016.

Feiring, Alice. *Naked Wine: Letting Grapes Do What Comes Naturally*. Cambridge, MA: Da Capo Press, 2011.

Fox, Michael V. *The Song of Songs and the Ancient Egyptian Love Songs*. Madison: University of Wisconsin Press, 1985.

Goode, Jamie. *The Science of Wine: From Vine to Glass, Second Edition*. Berkeley: University of California Press, 2014.

Goode, Jamie, and Sam Harrop. *Authentic Wine: Toward Natural and Sustainable Winemaking*. Berkeley: University of California Press, 2011.

Heekin, Deirdre. *Libation, A Bitter Alchemy*. White River Junction, VT: Chelsea Green Publishing, 2009.

Heekin, Deirdre. *An Unlikely Vineyard: The Education of a Farmer and Her Quest for Terroir*. White River Junction, VT: Chelsea Green Publishing, 2014.

Ibn al-Farid, Umar. *Sufi Verse, Saintly Life*. Translated by Emil Homerin, Mahwah, NJ: Paulist Press, 2001.

Lefkowitz, Mary, and Maureen Fant. *Women's Life in Greece and Rome: A Source Book in Translation*. London: Bloomsbury Academic, 2016.

Levy, Avigdor, editor. *Jews, Turks, Ottomans: A Shared History, Fifteenth Through the Twentieth Century*. Syracuse, NY: Syracuse University Press, 2002

Lichtheim, Miriam. *Ancient Egyptian Literature: A Book of Readings, Vol. I, The Old and Middle Kingdoms*. Berkeley: University of California Press, 1973.

Lukacs, Paul. *Inventing Wine: A New History of One of the World's Most Ancient Pleasures*. New York: W.W. Norton & Company, 2012.

Lynch, Kermit. *Adventures on the Wine Route: A Wine Buyer's Tour of France*. New York: Farrar, Straus and Giroux, 1988.

McGovern, Patrick. *Ancient Wine: The Search for the Origins of Viniculture*.

Princeton, NJ: Princeton University Press, 2003.

McGovern, Patrick. *Uncorking the Past: The Quest for Wine, Beer, and Other Alcoholic Beverages.* Berkeley: University of California Press, 2010.

McGovern, Patrick E., Stuart J. Fleming, and Solomon H. Katz, editors. *The Origins and Ancient History of Wine.* London: Routledge, 1996.

McMullen, Thomas. *Hand-Book of Wines, Practical, Theoretical, and Historical: With a Description of Foreign Spirits and Liqueurs.* New York: D. Appleton and Company, 1852.

Mühifried, Florian. *Being a State and States of Being in Highland Georgia.* New York: Berghahan Books, 2014.

Phillips, Rod. *Alcohol: A History.* Chapel Hill: University of North Carolina Press, 2014.

Phillips, Rod. *French Wine: A History.* Oakland: University of California Press, 2016.

Patton, Rev. William. *Bible Wines: Or, The Laws of Fermentation and Wines of the Ancients.* New York: National Temperance Society and Publication House, 1874.

Pollan, Michael. *The Botany of Desire: A Plant's Eye View of the World.* New York: Random House, 2001.

Plato. *Laws.* Translated by Benjamin Jowett. Oxford University: Clarendon Press, 1875.

Robinson, Jancis, Julia Harding, and José Vouillamoz. *Wine Grapes: A Complete Guide to 1,368 Vine Varieties, Including Their Origins and Flavours.* New York: Ecco Press, 2012.

Robinson, Jancis, editor. *The Oxford Companion to Wine, Third Edition.* New York: Oxford University Press, 2006.

Robinson, Jancis, editor. *The Oxford Companion to Wine, Fourth Edition.* New York: Oxford University Press, 2015.

Scheindlin, Raymond P., translator. "A Miniature Anthology of Medieval Hebrew Wine Songs," *Prooftexts* 4, no. 3, Sept. 1984, Indiana University: 269–300.

Scheindlin, Raymond. *Wine, Women, and Death: Medieval Hebrew Poems on the Good Life.* New York: Oxford University Press, 1999.

Shepherd, Gordon M. *Neuroenology: How the Brain Creates the Taste of Wine.* New York: Columbia University Press, 2017.

Smith, Clark. *Postmodern Winemaking: Rethinking the Modern Science of an Ancient Craft.* Berkeley: University of California Press, 2014.

Washington, Peter, editor. *Persian Poets.* New York: Everyman's Library, Alfred A. Knopf, 2000.

Wohlleben, Peter. *The Hidden Life of Trees: What They Feel, How They Communicate—Discoveries from a Secret World.* Translated by Jane Billinghurst. Vancouver: Greystone Books, 2015.

TASTING THE PAST

Tasting the Past—and the Future:
An Update from Kevin Begos

The Hidden Life of Wine:
An Interview with Kevin Begos

TASTING THE PAST—AND THE FUTURE

An Update from Kevin Begos

A realization slowly emerged as I walked through rock-strewn Washington vineyards with Kevin Pogue in early winter. We were wine-geeking about terroir, a term that continues to spawn endless debates over exactly how or why vines grown in particular soils and rocks sometimes have unique minerally flavors.

Pogue teaches geology at Whitman College in Walla Walla and says things like, "There's a big chunk of granite. Look at that. That's beautiful." Spending a day with Pogue was like having John McPhee give a vineyard tour based on his classic geology book *Basin and Range*—the rocks came to life. He explained how an enormous glacial lake formed and burst repeatedly many thousands of years ago, up until the last Ice Age. The floods covered thousands of square miles and reached from the mountains to the Pacific Ocean, carrying untold numbers of boulders and rocks as if they were pebbles, and sweeping vast flows of sediment across plains. Some scientists believe the largest glacial lake bursts had the energy equivalent of 4,500 megatons of TNT. The Hiroshima atomic bomb was only a tiny fraction of one megaton.

The end result of the cataclysms was that some parts of Washington State are entirely covered with fine, silty soil that goes down hundreds of feet, others have a thin top layer of silt with different rocky soil underneath, and some are literally fields of golf- and bowling-ball-sized rocks that extend down fifty or a hundred feet.

The soil differences affect how vines grow, and also how people perceive wines. That's because minerally tastes are especially fashionable now among critics—Alice Feiring even has a new book titled *The Dirty Guide to Wine*. Rocky vineyards and mineral flavors are also associated with higher-quality wines, so there has been an explosion of labels that evoke rocks.

The shelves of Pogue's office are, of course, filled with beautiful rocks, but he also has a label collection that has grown to cover two walls, featuring wines called Schist, Basalt, and Syncline. The last is odd because Merriam-Webster defines that as "a trough of stratified rock in which the beds dip toward each other." So are we supposed to be able to taste rock formations in wine, too? At that point terroir becomes the viticultural version of Santa Claus—a seductive concept that shouldn't be taken too literally.

But Pogue is trying to bring some real science to the debate. He's done research that led to the only American Viticultural Areas (AVA) with boundaries that directly correspond to the soils and rocks below. France and other nations have specific viticultural areas where only certain grapes are allowed to grow—think Bordeaux and Burgundy—but the boundaries were drawn by merchants and winemakers, not geologists.

As I consider wine trends in 2019 the visit with Pogue resonates, because it reminded me of how wine science and wine tradition are increasingly colliding in both pleasing and unsettling ways. The Washington geologic AVAs actually gave us large-scale field trials of how different soils and rocks affect the taste of wines made from the same grape (in this case usually Syrah).

Exciting things are also happening with hybrid grapes in small vineyards around the world. Bordeaux's Ducourt vineyards planted vines that contain disease-resistant genes, and German vineyards have done the same. The US Department of Agriculture is funding research called VitisGen—the vineyard equivalent of the Human Genome Project. It is an attempt to use the vast power and rapidly declining cost of DNA research to pinpoint the precise chromosomal locations in wild American grapes that drive flavors, aromas, grape size, and other important attributes. VitisGen now includes scientists from Cornell, the University of California at Davis, the University of Minnesota, and seven other universities. Industry giant E. & J. Gallo is also a partner.

Yet tradition resists change in other ways. New American hybrids that cross famous grapes with wild ones clearly have potential, but one expert pointed to challenges. Geoff Kruth, a master sommelier and the president of GuildSomm, an international nonprofit based in California, wrote me in an email that for US consumers grape variety and wine style are strongly linked, and "it takes quite a bit of time and exposure for new grapes to catch on with the drinking public. I think there is an increasing openness to unfamiliar varieties, but in the end it will be driven by both quality and availability. If the quality is there, unknown varieties with good yields can always find a home in blends or niche bottlings, but you wouldn't want to be in a position to have to sell large quantities of any wine without a familiar grape variety or brand-name blend."

Not only that, money is still pouring into wine and spirits start-ups that have radically different ideas about the future. In 2018 *Food & Wine* reported that Ava Winery, which had sought to make synthetic wine without using grapes, has put that plan on hold and switched to producing "molecular whiskey" under the name Endless West.

A *Wall Street Journal* reporter tried a sample and felt "it seemed to be missing some ineffable, essential whiskey quality—I couldn't put my finger on exactly what—but I liked it." After the rebrand, Endless West raised $10 million in new seed money and "hired a small staff of food scientists and analytical chemists." They still hope to produce synthetic wine someday.

Other barriers to change are distinctly old-fashioned. The critic Jon Bonné wrote an excellent and widely debated 2018 essay, "Why Is the Wine World So Un-Woke?" It discussed blind spots in gender, class snobbery, labor issues, and other topics, and observed that the industry "has a nearly pathological aversion to its less-than-perfect side."

Bonné suggested the Anthony Bourdain spirit instead: a "willingness to poke at the dark underlining of putatively shiny things—including the food world he loved . . . He felt that an essential part of loving a thing was to tear down the false romanticism surrounding it."

Pesticide is one word Bonné didn't mention. Yet in California alone more than 260 million pounds were applied to vineyards between 2007 and 2016, according to state records. "The long-term trend over the last two decades is an increasing area treated for all pesticide types except for sulfur, which has tended to fluctuate," officials note.

But scientists are zeroing in on precise spots in wild grape DNA, which contains natural resistance to pests and diseases. Faced with the long-term choice between massive pesticide applications and a little bit of cross-breeding (a.k.a. letting famous grapes have sex with obscure ones), why do we even hesitate? Perhaps because we haven't really had the uncomfortable debate about who is most at risk: not the wine drinker or the winemaker, but vineyard workers and nearby communities. Pesticides can be sprayed ten or fifteen times yearly. In France there have already been lawsuits over worker exposure, and a documentary about risks to the general public, including schoolchildren. I think wine writers and sommeliers should recognize that dark side of vineyards, and promote scientific options that can limit pesticide use.

Many wine critics continue to praise rare and unusual grapes, but an influential New York City sommelier also exposed what small quality producers are up against. Victoria James, author of *Drink Pink: A Celebration of Rosé*, described the shady marketing offers she has refused in a recent article for *Bon Appétit*. "The deal was that they would give me a couple thousand in cash to be an ambassador, and I would have to buy their rosé to pour by the glass for the summer." Instead she started the campaigns #sommsforrealrosé and #nowayshittyrosé.

I cringed while reading James: she notes many mass-market bottles of rosé "come from huge swaths of land, particularly in California and Provence, with 'terroir' barely suitable for even vegetables. Bulk wines—and there are hundreds of them—are owned by large companies with deep pockets, with big marketing budgets. Money is channeled away from the high-quality grape production and toward massive advertising campaigns coupled with paid inclusion on hot restaurant menus."

Still another bright spot is that the industry is beginning to take climate change threats seriously. Al Gore is scheduled to be the keynote speaker at a 2019 conference in Portugal on climate and winemaking, along with Swiss scientist José Vouillamoz and experts from France, Spain, England, and America.

Climate change is a serious threat to traditional winemaking areas, and a 2018 paper in *Nature Climate Change* (Wolkovich et al.) found that heat-resistant, late-maturing varieties are still "almost entirely excluded" from many new plantings, and winemakers are "investing in an increasingly limited portfolio at exactly the time when a large diversity of varieties is most needed."

The biggest news? Wines made from grapes I'd never heard of only a decade ago are easier and easier to find, and more grapes are being rediscovered each year. My *Tasting the Past* wish list now includes wines from Portugal (Rabigato and Jampal), Austria and Hungary (Zierfandler and Ezerjó), and Spain (Garró and Tinto Jeromo).

Wherever you live or travel, seek out unusual wines. They are often delicious and always thought provoking. Over the last year I've met people all over the country who love the stories of rare and ancient grapes, so you can probably find someone to share a bottle with.

Kevin Begos
June 2019

THE HIDDEN LIFE OF WINE

An Interview with Kevin Begos

Malcolm Jolley, a founding editor of Good Food Revolution *and executive director of Good Food Media, talks with Kevin Begos.*

Good Food Revolution: I really enjoyed your book; it's fantastic. I know some of the people in it, and have been to a few of the places that you go to, like Georgia and Sicily, and tried the indigenous grapes. Let's talk about that history a little bit. Certainly if you go to Georgia they'll let you know that they're convinced that this is where wine was invented, and you better not bring up that the Armenians and the Persians also have a claim on it. But, in any event, the general idea is that the grapes that human beings grow to make wine came from the Caucasus and then migrated westerly across the Mediterranean, right?

Kevin Begos: That's right. Both westerly and then also northward up through Europe, often through the river valleys of the Danube, the Rhône, and the Rhine. You can actually trace the DNA, and that part of the story fascinated me. It's just like what you can do with the genealogy of your own family—the DNA of grapes has a genealogy and it shows where different varieties split off and combined with local vines, which ones are older and which ones are younger. So the Georgians do, in fact, have some DNA evidence to back up their claims. It's not just their pride in the region.

GFR: The publication of *Tasting the Past* is good timing, then, in the sense that we're just at the point where we can map grape genome. At the same time, there is a growing interest in indigenous grapes—certainly among growers and winemakers in Europe.

KB: You know it's definitely a growing movement from people all over Europe and the Middle East, that's now going back to ten or sometimes even fifteen years ago. And the *Wine Grapes* book by Julia Harding, Jancis Robinson, and José Vouillamoz really helped when it came out in 2012, listing over 1,300 varieties. But I think the movement toward indigenous grapes is just starting here in America. Of course, there's the slow food movement that has been here for years. And there is the craft beer and the craft distilleries movements, but winemakers here I think are a little late in realizing that there may be a lot of possibilities with our hybrid grapes that use some of the Native American DNA crossbred with European. Of course, that is still contentious.

GFR: Well, we certainly still make hybrid grape wines here in Ontario and actually some of them make quite decent wines. But they can also be a bit of an acquired taste, if you know what I mean.

KB: Sure, I felt that way for a long time, but Deirdre Heekin's wine from Vermont kind of changed my mind. She's using some of these hybrid grapes that came out of the University of Minnesota, cold hardy hybrids. I think it speaks to the importance of the winemaker's skill, that she did change my mind. Deirdre spent a lot of time in Northern Italy learning about wine culture and food culture there.

Her wine, made of these hybrid grapes, I think in most blind tastings you might not guess where it's from, but you wouldn't say Vermont. You know, it's really a classy wine and a number of other wine critics agree: tough and experienced people like Eric Asimov from the *New York Times* and the writer Alice Feiring.

GFR: Okay, but it's not all just grapes and winemakers. You have a chapter on that elusive concept of terroir—how does that all fit in?

KB: Well, we know clearly that terroir is a lot more than just the soil, but that's not to say the soil isn't important. I mean of course good soil is important for any vine or fruit or vegetable. Any of us that have a garden know that. But the flavors in wine clearly don't come directly from the soil—what we call the mineral flavors. There's even some early evidence that some of the flintlike flavors in wines come from yeast and the fermentation process. And now we know the whole microbial community in the soil also clearly makes a difference to taste. I quote one of the California studies that found unique microbial environments in different California wine regions that affect how the vines grow. That's a big part of it because the grape genome has tremendous flavor resources, but how those genes express those flavors (or not) is going to come out of the local growing conditions, as well as the winemaker's skill.

GFR: One of the things I like about your book is that at the end of every chapter you have some information about where to find some of the wines you've written about, or even other producers to try. What about you? Are you drinking wine in a different way than you did before you started writing *Tasting the Past*?

KB: Oh, completely. I confess I had been a French, Italian, and some Spanish lover for all my life until five or ten years ago. This has just expanded my horizons and I'm embarrassed it took me so long for it to happen. But, you know, we just didn't have the same wine traditions here in America. Italy always had its native wine regions and you could go to the different parts of Sicily or Northern Italy and try different wines made from different grapes. But that's a fairly new idea for a lot of American wine drinkers. I think the sommeliers, the really trained professionals, have known about the other grapes for a while but the public is just catching on. There are so many types of wines to experience, like the qvevri wines from Georgia. That was a real revelation to me. You know that all those early wines for the first three, four, or five thousand years were made in a completely different style; not in oak barrels, not in barrels at all.

GFR: That brings up an interesting question about technology. Didn't the Romans in France just start using oak because they had it? Should modern winemakers use whatever technology is at hand too?

KB: I try to make the point that we've influenced grape evolution pretty heavily for eight thousand years by increasing berry size and choosing particular flavors. But now we're entering a whole new era, with the growth of DNA research and mass spectrometry tools that provide great stories and great information about the genome and what people drank in the past. They also provide tools to make synthetic wine or to tweak wine industrially with all sorts of things. I'm not a big fan of that. It's going to clearly be a battle that plays out over the next few decades with, I think, increasing intensity. The technology is going to get cheaper and cheaper and faster and faster. But yet, as you know from the good food movement, even with this, you know, super-connected crazy world we live in, more and more people are looking back toward slow food and sustainable agriculture. It may not be the majority of people but there is definitely a global interest in it. It's saying food doesn't have to be at hyperspeed.

GFR: Right. Absolutely. I think you make the point in the book that it is a richer world to have more of these interesting heritage grapes being made into wines. And it seems like you had a lot of fun researching the book. At least that's how I read it.

KB: Yes, it was fun. There was a constant discovery of surprises. I really had no idea that it was going to be so interesting in so many different directions. Toward the end of the book there's a quote from Yotam Ottolenghi, the Israeli-British chef, about the importance of keeping the diversity of flavor. It's just like heirloom vegetables. So, I think keeping grape diversity is the right thing to do. It's not like we're going to end up with inferior wines, or stop drinking French and Italian wines. We just need to open ourselves up to more possibilities.

First published in **Good Food Revolution,** *where it was edited for length, clarity, and style. Reprinted by permission.* GoodFoodRevolution.com

KEVIN BEGOS is a former MIT Knight Science Journalism fellow and a former AP correspondent whose stories and research have appeared in major newspapers and other publications, including *Scientific American, Harper's, Salon,* the *Christian Science Monitor,* the *Guild of Sommeliers, USA Today,* and the *New York Times.* His website is kevinbegos.com.